Deforestation Trends in
the Congo Basin

Deforestation Trends in the Congo Basin

Reconciling Economic Growth and Forest Protection

Carole Megevand

with Aline Mosnier, Joël Hourticq,
Klas Sanders, Nina Doetinchem,
and Charlotte Streck

THE WORLD BANK
Washington, D.C.

Contents

Boxes

Figures

Map

Tables

Foreword

As global networks of trade, migration, information, and finance have grown in strength, speed, and density over the last decades, so have our understanding and awareness of the connections that shape the world's physical landscapes and economies. We know that policy decisions in one country can affect the way land is managed thousands of miles away. We know that greenhouse gases emitted in different sectors of the economy and in different countries influence the pace of climate change for all. And we know that vicious cycles of poverty, land degradation, and food insecurity can be transformed into virtuous cycles of sustainable intensification and shared prosperity with the right interventions and incentives. Development challenges and solutions are all connected at the local, regional, and global level.

Those far-reaching connections come to the fore in a new and timely study that looks at deforestation trends in the Congo Basin across sectors and beyond national borders. The study, led by the World Bank's Africa Region Environment team with the participation of the key Congo Basin country stakeholders and support from multiple donors, was informed by economic modeling complemented with sectoral analysis as well as interactive simulations and workshop discussions. This innovative approach has already deepened our understanding of the multiple drivers of deforestation in the Congo Basin beyond the usual suspects (commercial logging) and opened political space to discuss the role of sectors such as agriculture, energy, transport, and mining in shaping the future of the Basin's forests.

This analysis, combined with recommendations that policy makers can now further refine and flesh out at the country level, could potentially help Congo Basin countries overcome some of the more severe trade-offs between growth and forest protection. If Congo Basin countries are able to minimize forest loss as their economies develop, they could "leapfrog" the steep drop in forest cover that has historically accompanied development in many countries, and make an important global contribution to climate change mitigation by reducing emissions associated with deforestation.

The time is now ripe to move ahead with some of the sound "no-regrets" recommendations made by study participants and experts.

Jamal Saghir
Director
Sustainable Development Department
Africa Region
The World Bank

Acknowledgments

This report was written by Carole Megevand with contributions from Aline Mosnier, Joël Hourticq, Klas Sanders, Nina Doetinchem, and Charlotte Streck. It is the output of a two-year exercise implemented at the request of the COMIFAC (Regional Commission in Charge of Forestry in Central Africa) to strengthen the understanding of the deforestation dynamics in the Congo Basin.

The modeling exercise was conducted by an IIASA team, led by Michael Obersteiner and composed of Aline Mosnier, Petr Havlík, and Kentaro Aoki. The data collection campaign in the six Congo Basin countries was coordinated by ONF-International under the supervision of Anne Martinet and Nicolas Grondard. The team is grateful to Raymond Mbitikon and Martin Tadoum (COMIFAC) as well as to Joseph Armaté Amougou (Cameroon), Igor Tola Kogadou (Central African Republic), Vincent Kasulu Seya Makonga (Democratic Republic of Congo), Deogracias Ikaka Nzami (Equatorial Guinea), Rodrigue Abourou Otogo (Gabon), and Georges Boudzanga (Republic of Congo) who helped put together a team of national experts who provided very useful insights and contributions used to fine-tune the modeling exercise.

The report drew on a series of background papers covering different sectors prepared by Carole Megevand jointly with Joël Hourticq and Eric Tollens on agriculture; with Klas Sanders and Hannah Behrendt on woodfuel energy; with Nina Doetinchem and Hari Dulal on forestry; with Loic Braune and Hari Dulal on transport; and with Edilene Pereira Gomes and Marta Miranda on mining. Charlotte Streck, Donna Lee, Leticia Guimaraes, and Campbell Moore prepared the section on the status of REDD+ in the Congo Basin and helped in preparing the consolidated report.

The team is grateful for the useful guidance provided by Kenneth Andrasko, Christian Berger, and Gotthard Walser. It also acknowledges constructive comments from Meike van Ginneken, Benoit Bosquet, Marjory-Anne Bromhead, Stephen Mink, Shanta Devaradjan, Quy-Toan Do, John Spears, Andre Aquino, Gerhard Dieterle, Peter Dewees, James Acworth, Emeran Serge Menang, Simon Rietbergen, David Campbell Gibson, Andrew Zakhrarenka, Loic Braune, Remi Pelon, and Mercedes Stickler.

The report was ably edited by Flore Martinant de Preneuf and Sheila Gagen. Maps and illustrative graphs were prepared by Hrishikesh Prakash Patel.

Special thanks go to Idah Pswarayi-Riddihough, Jamal Saghir, Ivan Rossignol, Giuseppe Topa, Mary Barton-Dock, and Gregor Binkert who, at different stages, helped make this initiative yield full results.

The study benefited from financial support from various donors, including DfID, Norway through the Norwegian Trust Fund for Private Sector and Infrastructure (NTF-PSI), the Program on Forests (PROFOR), the Trust Fund for Environmentally and Socially Sustainable Development (TFESSD), and the Forest Carbon Partnership Facility (FCPF).

About the Author and Contributors

Carole Megevand has 15 years of professional experience in Natural Resources Management (NRM) in developing countries. She holds two master's degrees, respectively, on Agricultural Economics and Environment/Natural Resources Economics. At the World Bank, she manages complex NRM operations in the Congo Basin countries with a specific focus on intersectoral dimensions and governance issues. For the past three years, she has been coordinated the REDD+ portfolio in the Africa Region for the World Bank. Her international experience in the developing world includes two long-term assignments (Cameroon and Tunisia) and missions in more than 15 countries in Africa, the Middle East and North Africa, and Latin America and the Caribbean.

Contributors

Aline Mosnier is a research scholar at the Ecosystems Services and Management (ESM) Program at IIASA. Her background is in development economics with a special focus on trade policies and rural development. Since 2008, she has contributed to the development of the GLOBIOM model—a global partial equilibrium model on land use change—especially on international trade, internal transportation costs, and biofuels aspects. In 2010, she was responsible for the adaptation of the GLOBIOM model to the Congo Basin context to provide estimates of future deforestation and support national REDD strategies in the region.

Joël Hourticq holds degrees from the Institut National Agronomique and the Ecole Nationale du Génie Rural des Eaux et Forêts in Paris and an MSc in agricultural economics from the University of London. He regularly collaborates with the World Bank, especially on agricultural public expenditure issues.

Klas Sander is a Natural Resource Economist with the World Bank's South Asia region. He has more than 13 years of professional experience in development with field experience in Africa, Southeast Asia, South Asia, West Asia, the Pacific, and Eastern Europe. Klas holds two master's degrees in Forestry and Agricultural Economics and a PhD in Rural Development. Klas' work focuses on natural resource economics and management, renewable energy, and economic and financial analyses. He recently conducted extensive analyses of wood-based energy systems and supply chains, including their potential for low-carbon development and green growth, with a particular focus on Africa.

Nina Doetinchem has been working on natural resources issues, in particular in the fisheries and forestry sectors, for over a decade. Raised in Kenya, she has since maintained a special interest in the African continent, its people, and its biodiversity. With an academic background in biology and environmental management and 10 years of experience in project implementation, she has focused her work on the often competing interface of commercial interests in natural resources, small-scale livelihoods, and conservation of biodiversity.

Charlotte Streck is an internationally recognized expert on the law and policy of climate change and emissions trading. She advises on the regulatory framework of international and national climate policy and is a leading expert on the consideration of forestry and agriculture in climate change regulation. Charlotte reads environmental law and diplomacy in frequent lectures and is a prolific writer and publisher. She serves as associate editor of *Climate Policy* and on the editorial board of several other academic journals. She is cofounder of the Global Public Policy Institute (Berlin) and Avoided Deforestation Partners (Berkeley).

Abbreviations

AICD	Africa Infrastructure Country Diagnostic
ARM	Alliance for Responsible Mining
ASTI	Agricultural Science and Technology Indicators
CAADP	Comprehensive Africa Agriculture Development Program
CBD	Convention on Biological Diversity
CBFP	Congo Basin Forest Partnership
CEMAC	Economic and Monetary Community of Central Africa
CGIAR	Consultative Group on International Agricultural Research
CIFOR	Center for International Forestry Research
COMIFAC	Forestry Commission of Central Africa
ECCAS	Economic Community for Central African States
EIAs	environmental impact assessments
EITI	Extractive Industries Transparency Initiative
EPIC	Environmental Policy Integrated Climate
EU ETS	European Emissions Trading System
FAO	Food and Agriculture Organization of the United Nations
FCFA	African Financial Community Franc
FCPF	Forest Carbon Partnership Facility
FIP	Forest Investment Program
FLEGT	Forest Law Enforcement, Governance, and Trade (European Union)
FT	forest transition
GDP	gross domestic product
GEF	Global Environment Facility
GHG	greenhouse gas
GIS	geographic information system
HFLD	high forest cover and low deforestation
IEA	International Energy Agency
IFPRI	International Food and Policy Research Institute

IIASA	International Institute for Applied Systems Analysis
IRAD	Institut de Recherche Agricole pour le Développement
ITTO	International Tropical Timber Organization
IUCN	International Union for Conservation of Nature
LICs	low-income countries
LPI	Logistic Performance Index
MINFOF-MINEP	Ministry of the Forests and Ministry of Environment (Cameroon)
MP	management plan
NEPAD	New Partnership for Africa's Development
NICFI	Norway International Climate and Forest Initiative
NGOs	nongovernmental organizations
NTFPs	nontimber forest products
OFAC	Observatory for the Forests of Central Africa (Observatoire des Forêts d'Afrique Centrale)
R&D	research and development
REDD+	Reducing Emissions from Deforestation and Forest Degradation Plus
SEZ	special economic zone
SFM	sustainable forest management
SIAs	social impact assessments
UNCBD	UN Convention on Biological Diversity
UNCCD	UN Convention to Combat Desertification
UNFCCC	United Nations Framework Convention on Climate Change
VPAs	voluntary partnership agreements
WWF	World Wildlife Fund

Overview

Congo Basin Forests at a Glance

The Congo Basin spans six countries: Cameroon, the Central African Republic, the Democratic Republic of Congo, the Republic of Congo, Equatorial Guinea, and Gabon. It contains about 70 percent of Africa's forest cover: Of the Congo Basin's 530 million hectares of land, 300 million are covered by forest. More than 99 percent of the forested area is primary or naturally regenerated forest as opposed to plantations, and 46 percent is lowland dense forest.

Industrial logging represents an extensive land use in the area, with about 44 million hectares of forest under concession (8.3 percent of the total land area), and contributes significantly to revenues and employment (figure O.1). Unlike other tropical regions, where logging activities usually entail a transition to another land use, logging in the Congo Basin is highly selective and extensive and production forests remain permanently forested.

The Congo Basin forests are home to about 30 million people and support livelihoods for more than 75 million people from over 150 ethnic groups who rely on local natural resources for food, nutritional health, and livelihood needs. These forests provide crucial protein sources to local people through bushmeat and fisheries. Forest products, whether directly consumed or traded for cash, provide a substantial portion of local people's income. Forests also constitute an important safety net in countries where poverty and undernourishment are highly prevalent (box O.1).

These forests perform valuable ecological services at local, regional, and global levels. Local and regional services include maintenance of the hydrological cycle and important flood control in a high-rainfall region. Other important regional benefits include regional-scale climate regulation, cooling through evapotranspiration, and buffering of climate variability. The forests also house an enormous wealth of plant and animal species including threatened animals such as the lowland gorilla and chimpanzee. Globally, Congo Basin forests represent about 25 percent of the total carbon stored in tropical forests worldwide, mitigating anthropogenic emissions (de Wasseige et al. 2012).

Figure O.1 Land, Dense Forest, and Logging Areas in the Congo Basin

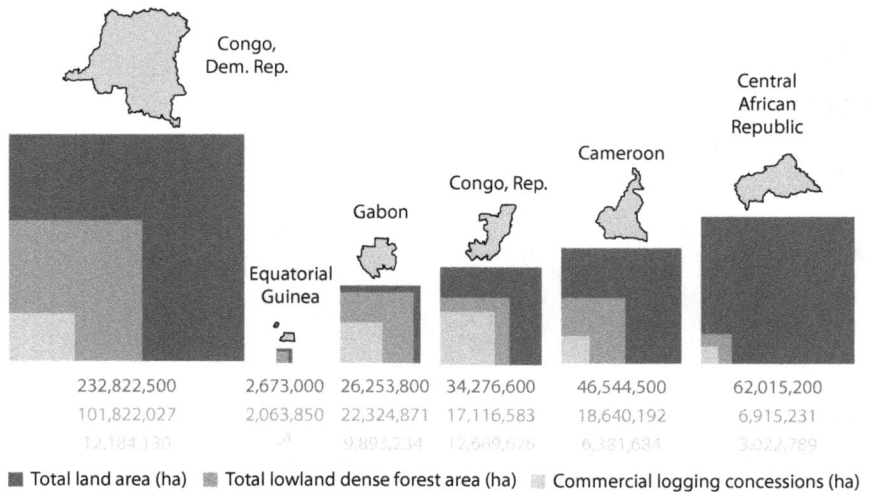

232,822,500	2,673,000	26,253,800	34,276,600	46,544,500	62,015,200
101,822,027	2,063,850	22,324,871	17,116,583	18,640,192	6,915,231
12,184,130	a	9,893,234	12,669,625	6,381,684	3,022,789

■ Total land area (ha)   ■ Total lowland dense forest area (ha)   ▧ Commercial logging concessions (ha)

Source: Data prepared from de Wasseige et al. 2012.
Note: ha = hectare.
a. In Equatorial Guinea, all logging concessions were cancelled in 2008.

Box O.1 Hunger in a Land of Plenty

Although most Congo Basin countries are richly endowed with natural resources and abundant rainfall, hunger is a serious to extremely alarming concern in all countries except Gabon (IFPRI Global Hunger Index 2011). Agriculture is still dominated by traditional low input and low output subsistence systems, and there are tremendous gaps between actual and potential yields. Poor infrastructure isolates farmers from potential markets and growth opportunities, thereby cutting off a significant proportion of the Congo Basin's population from the broader economy.

Table O.1.1 Key Development Indicators for Congo Basin Countries

	Poverty	Nutrition	Agricultural land	Employment	Access to food
Country	Population at purchasing power parity US$1.25 a day (%)	Children under age five underweight (%)	Agricultural land area (% of total land area)	Economically active population in agriculture (%)	Paved roads (% of total roads)
Cameroon	9.6	16.6	19.8	46.4	8.4
Central African Republic	62.8	21.8	8.4	62.3	…
Congo, Dem. Rep.	59.2	28.2	9.9	56.7	1.8
Congo, Rep.	54.1	11.8	30.9	31.2	7.1
Equatorial Guinea	…	10.6	10.9	63.8	…
Gabon	4.8	8.8	19.9	25.2	10.2
Sub-Saharan Average	**47.5**	**21.3**	**52.6**	**58.2**	**23.8**

Source: United Nations Development Programme (UNDP) 2012.

Figure O.2 Average Annual Net Deforestation and Net Degradation Rates, Congo Basin, 1990–2000 and 2000–05

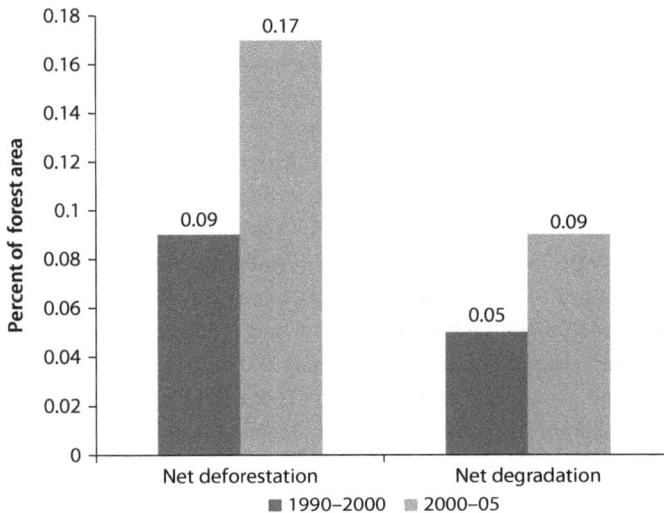

Source: de Wasseige et al. 2012.

Deforestation and forest degradation have been minimal in the Congo Basin. Africa as a whole is estimated to have contributed only 5.4 percent to the global loss of humid tropical forest during 2000–05, compared to 12.8 percent in Indonesia and 47.8 percent for Brazil alone (Hansen et al. 2008). However, deforestation in the Congo Basin has accelerated in recent years (figure O.2). Deforestation and forest degradation have been largely associated with expansion of subsistence activities (agriculture and energy) and concentrated around densely populated areas.

What Will Drive Deforestation in the Congo Basin? A Multisectoral Analysis

The Congo Basin forests may be at a turning point, heading toward higher deforestation and forest degradation rates. The Congo Basin forests have been mainly "passively" protected by chronic political instability and conflict, poor infrastructure, and poor governance. Congo Basin countries thus still fit the profile of high forest cover and low deforestation (HFLD) countries. However, there are signs that Congo Basin forests are under increasing pressure from a variety of sources, including mineral extraction, road development, agribusiness, and biofuels, in addition to subsistence agricultural expansion and charcoal collection.

Causes and drivers of tropical deforestation are complex and cannot easily be reduced to a few variables. The interplay of several proximate as well as underlying factors drives deforestation in a synergetic way. Expansion of subsistence activities (agriculture and woodfuel collection) is the most commonly cited proximate cause of deforestation in the Congo Basin. This is supported by demographic

Box O.2 An Interactive Modeling and Awareness-Raising Exercise

In 2009, the six Congo Basin countries, along with donors and partner organizations, agreed to collaborate to analyze major drivers of deforestation and forest degradation in the region. A modeling approach was chosen because the high forest cover and low deforestation (HFLD) profile of the Congo Basin countries justified using a prospective analysis to forecast deforestation as historical trends were considered inadequate to capture the future nature and amplitude of drivers of deforestation. The approach built on an adaptation of the GLOBIOM model set up by the International Institute for Applied Systems Analysis (IIASA) and tailored to the Congo region (CongoBIOM) to investigate drivers of deforestation and resulting greenhouse gas emissions by 2030. It also strongly relied on inputs from three regional multistakeholder workshops held in Kinshasa and Douala, respectively, in 2009 and in 2010 as well as in-depth analysis of trends in the agriculture, logging, energy, transport, and mining sectors.

The CongoBIOM was used to assess the impacts of a series of "policy shocks" identified by the Congo Basin country representatives. Various scenarios were developed to highlight both the internal and external drivers of deforestation:

- **Internal drivers**: improved transport infrastructure, improved agricultural technologies, and decrease in woodfuel consumption.
- **External drivers**: increase in international demand for meat, and increase in international demand for biofuel.

trends and accelerated urbanization, which form the most important underlying cause of current deforestation. The Congo Basin region has so far not witnessed the expansion of large-scale plantation experienced in other tropical regions; however, larger macroeconomic trends could change this situation (box O.2).

Agriculture

Agriculture is a vital yet neglected sector in the Congo Basin. Agriculture remains by far the region's largest employer. In Cameroon, the Central African Republic, the Democratic Republic of Congo, and Equatorial Guinea, more than half of the economically active population is still engaged in agricultural activities. Agriculture is also a significant contributor to gross domestic product (GDP), particularly in Cameroon, the Central African Republic, and the Democratic Republic of Congo. Despite its importance, the agriculture sector has so far been neglected and underfunded for much of the past few decades. Most agriculture is small scale, and the sector is dominated by traditional subsistence systems with a few large commercial enterprises, focused mainly on palm oil and rubber. Agricultural productivity in the region is very low compared with other tropical countries, with overall very low fertilizer use. As a result, reliance on food imports is substantial and increasing.

The potential for agricultural development in the Congo Basin is significant for several reasons. First, Congo Basin countries are endowed with much suitable and available land: Altogether, Congo Basin countries represent about 40 percent of the noncultivated, nonprotected low-population-density land suitable for cultivation in Sub-Saharan Africa and 12 percent of the land available worldwide; if only nonforested suitable areas are included, the Congo Basin still represents about 20 percent of the land available for agricultural expansion in Sub-Saharan Africa and 9 percent worldwide (Deininger et al. 2011). Second, the region has unconstrained water resources, which gives it an edge over other areas that may face water scarcity as a result of climate change. Third, Congo Basin countries unsurprisingly rank among the countries with the greatest potential in the world for increasing yields. Finally, the rapidly urbanizing populations as well as increasing international demands for food and energy could drive a dramatic demand for agricultural products from the Congo Basin. These factors combine to make the agriculture a very promising sector.

Future agricultural developments may, however, be at the expense of forests. Unlocking the agriculture potential in the Congo Basin could increase pressure on forests, particularly if investments in road infrastructure remove a long-lasting bottleneck to market access. The CongoBIOM was used to identify the potential impacts of specific changes, both internal (such as agricultural productivity) and external (international demand for meat or palm oil) on Congo Basin forests. It highlights that the increase in agricultural productivity, often seen as a win–win solution to increase production and reduce pressure on forests, could turn out to drive more deforestation (box O.3).

Despite its marginal contribution to global markets, the Congo Basin could be affected by global trends in agricultural commodity trade. The CongoBIOM tested two scenarios dealing with international commodity demand:

- Scenario 1: increase in global meat demand by 15 percent by 2030.
- Scenario 2: doubling of first generation biofuel production by 2030.

Box O.3 Why Agricultural Productivity Increases Are Not Necessarily Good for Forests

An increase in agricultural productivity is often seen as the most promising means to address the challenge of both producing more food and conserving more forest to preserve vital ecosystem services. It is commonly assumed that producing more on the same amount of land should prevent the need to expand into the forest frontier and thus help preserve forests.

However, models show that this is unlikely to happen, unless the right accompanying measures are put in place. The CongoBIOM suggests that intensification of land production in the Congo Basin will lead to an expansion of agricultural land owing to a context of growing demand for food and unconstrained labor supply. Productivity gains that make agricultural activities more profitable tend to increase pressure on forested land, which is generally the easiest and cheapest land for farmers to access. Environmental degradation, land tenure, and customary rights issues associated with large-scale farmland acquisition are other factors driving farmers into forested land. Chapter 3 of the report outlines some of the countervailing policies and land planning measures that could help mitigate the impact of agricultural development on forests.

Under both scenarios, the CongoBIOM highlights that the Congo Basin is unlikely to become a large-scale producer of meat or biofuel (in the short or medium term), but that it will be indirectly affected by changes in other parts of the world.

As an example, although meat production in the Congo Basin is hampered by the prevalence of the tsetse fly and the absence of a reliable feedstock industry, it could still be affected by global increase in meat demand. According to the CongoBIOM model, an increase in meat production (associated with increased land areas devoted to pasture and feed crops) in other regions of the world would reduce the production of other crops traditionally imported by the Congo Basin countries (for example, maize); this would trigger a substitution of imports by more locally grown products that could potentially lead to increased deforestation in the Congo Basin (figure O.3).

Figure O.3 Channels of Transmission of Increase in Global Demand for Meat and Increase in Deforestation in Congo Basin

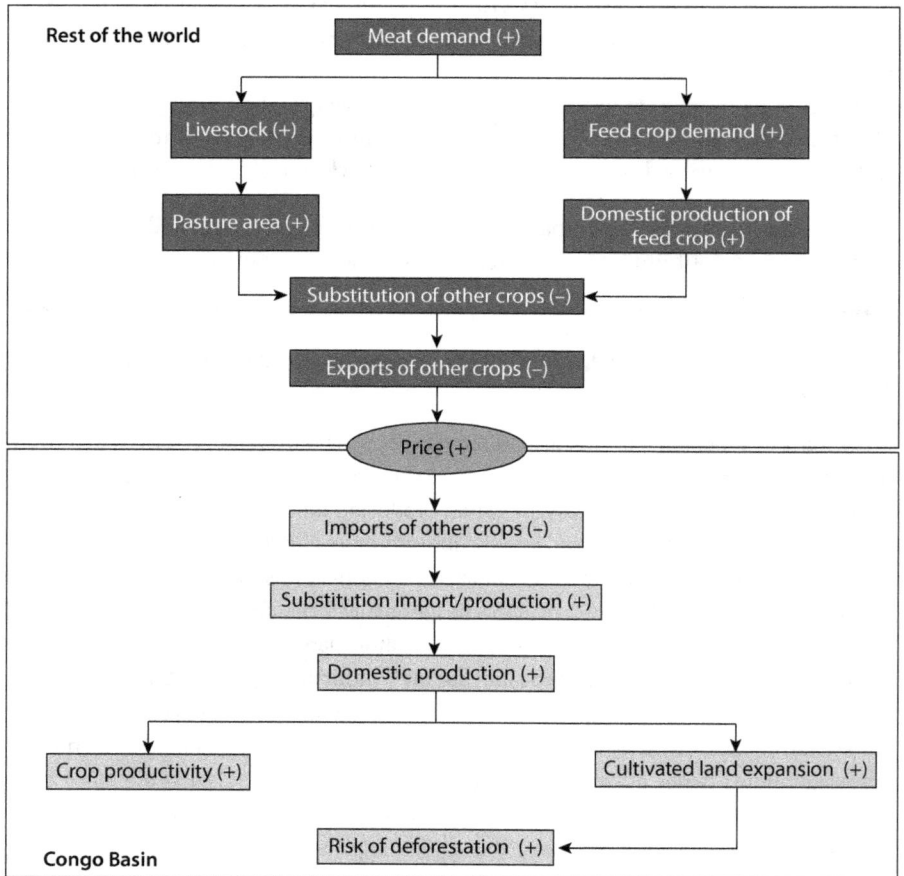

Note: + indicates increases; − indicates decreases.

Energy

It is estimated that more than 90 percent of the total volume of wood harvested in the Congo Basin is for woodfuel and that on average 1 cubic meter of woodfuel is required per person per year (Marien 2009). In 2007, the total production of woodfuel by Congo Basin countries exceeded 100 million cubic meters. The biggest producers were the Democratic Republic of Congo, with 71 percent of total regional woodfuel production, and Cameroon with 21 percent, reflecting the countries' shares of the regional population.

That said, energy profiles vary from one country to another based on wealth, access to electricity, and the relative costs of wood and fossil fuel energy. In Gabon, for example, the reliance on woodfuel is significantly lower, thanks to an extensive electricity network and subsidized gas for cooking.

The urban lifestyle tends to be more energy intensive because households in urban areas tend to be smaller leading to less efficient per capita fuel use for cooking. In addition, charcoal is often the primary cooking fuel for many small-scale roadside restaurants and in kitchens of larger public institutions, such as schools and universities, hospitals, and prisons, as well as small-scale industries. With an average urban growth of 3–5 percent per year and even higher (5–8 percent) for large cities such as Kinshasa, Kisangani, Brazzaville Pointe Noire, Libreville, Franceville, Port Gentil, Douala, Yaounde, and Bata, the Congo Basin countries are witnessing a shift from woodfuel to charcoal consumption because charcoal is cheaper and easier to transport and store.

Charcoal production in the Congo Basin increased by about 20 percent between 1990 and 2009.[1] In China, India, and much of the developing world, the use of wood-based biomass energy has peaked or is expected to peak in the near future. In contrast, in the Congo Basin (as well as in Africa in general), consumption of wood-based biomass energy is likely to remain at very high levels and even continue to increase for the next few decades; this trend is based on population growth, urbanization, and the relative price change of alternative energy sources for cooking such as liquefied petroleum gas or others (figure O.4).

Woodfuel collection becomes a serious threat to forests in densely populated areas. In rural areas, the impact of woodfuel collection may be offset by natural forest regeneration, but it can become a severe cause of forest degradation and eventual deforestation when demand comes from concentrated markets such as urban households and businesses. Basins that supply a growing urban demand extend over time and can radiate as far as 200 kilometers from the city centers, gradually degrading natural forests. The periurban area within a radius of 50 kilometers of Kinshasa, for example, has been largely deforested (box O.4).

Wood biomass energy is supplied by an inefficient sector. Charcoal is mostly produced using traditional techniques, with low transformation efficiencies (earth pit or earth mound kilns). The organization of the charcoal supply chain is also notoriously inefficient, relying on poorly designed regulatory frameworks that eventually lead to massive informality in the sector. The pricing structure of

Figure O.4 Number of People Relying on the Traditional Use of Biomass
millions

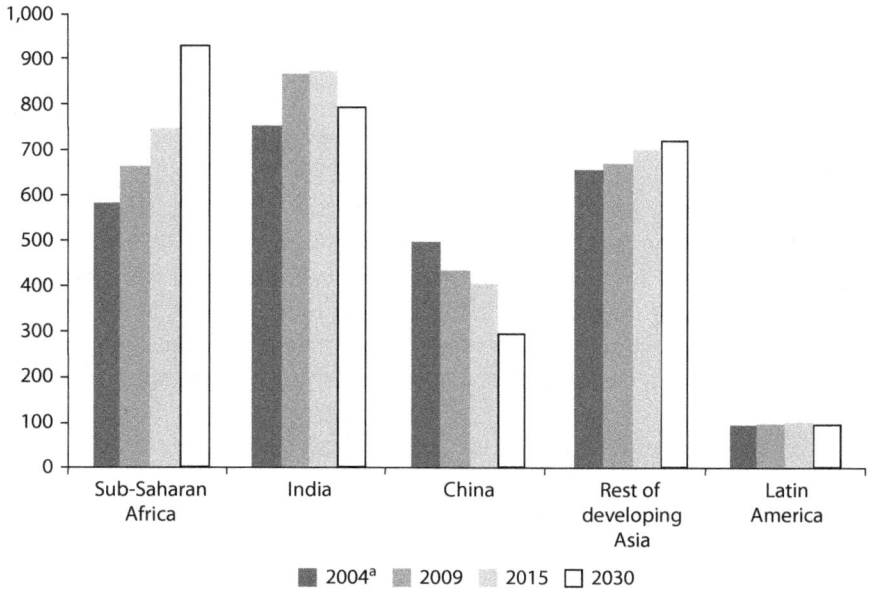

Source: International Energy Agency (IEA) 2010.
Note: The projections for 2015 and 2030 are part of the IEA "New Policies Scenario," which assumes that recent government commitments are implemented in a cautious manner, and primary energy demand increases by one-third between 2010 and 2035, with 90 percent of the growth in economies outside of those in the Organisation for Economic Co-operation and Development.
a. IEA World Energy Outlook 2006.

woodfuel sends perverse signals, as it incompletely accounts for the different costs along the value chain. In most cases, the primary resource (wood) is taken as a "free" resource. Inadequate economic signals in the woodfuel supply chain do not allow the producer to apply sustainable forest management (SFM) techniques.

However, experience in other countries (for example, Rwanda) suggests that the scarcity of wood products increases the economic value of remaining forests, thereby creating incentives for better forest management and the establishment of woodlots and tree plantations. As a result, forest ecosystems begin to recover—albeit with a great loss in biodiversity—and transition to more artificially planned plantations and monocultures.

Transportation

Transportation infrastructure in the Congo Basin is woefully inadequate to support development and poverty reduction. Road networks are sparse and poorly maintained, often as a result of recent civil conflicts. The paved road density in the Congo Basin is among the lowest in the world with only 25 kilometers of paved road for each 1,000 square kilometers of arable land, compared with an average of 100 kilometers in the rest of Sub-Saharan Africa. A legacy of

Box 0.4 Feeding Cities: Mixing Charcoal and Cassava near Kinshasa

Kinshasa, a megacity of 8–10 million inhabitants, is located in a forest–savanna mosaic environment on the Batéké Plateau in the Democratic Republic of Congo. The city's wood energy supply of about 5 million cubic meters per year is mostly informally harvested from degraded forest galleries within a radius of 200 kilometers off Kinshasa. With gallery forests most affected by degradation from wood harvesting, even forests beyond the 200 kilometers radius are experiencing gradual degradation, while the periurban area within a radius of 50 kilometers of Kinshasa has suffered total deforestation.

However, there have been several attempts to develop plantations around the megacity to help provide wood energy on a more sustainable basis. About 8,000 hectares of plantations were established in the late 1980s and early 1990s in Mampu, in the degraded savanna grasslands 140 kilometers from Kinshasa to meet the city's charcoal needs. Today, the plantation is managed in 25-hectares plots by 300 households in a crop rotation that takes advantage of acacia trees' nitrogen-fixing properties and the residue from charcoal production to increase crop yields. The plantations produce about 10,000 tons of charcoal per year, or 1.6 percent of Kinshasa's estimated charcoal demand (Peltier et al. 2010).

Another scheme, run by a Congolese private company called Novacel, intercrops cassava with acacia trees in order to generate food and sustainable charcoal, as well as carbon credits. To date, about 1,500 hectares have been planted. The trees are not yet mature enough to produce charcoal, but cassava has been harvested, processed, and sold for several years. The company has also received some initial carbon payments. The project has been producing about 45 tons of cassava tubers per week and generates 30 full-time jobs, plus 200 seasonal jobs. Novacel reinvests part of its revenue from carbon credits into local social services including the maintenance of an elementary school and health clinic.

the colonial era, the railways system was designed to facilitate the extraction of natural resources rather than to support the movement of people and goods. Railways are poorly maintained, with more than a third of the total network not fully operational. The river transportation networks of the Congo Basin hold great potential (25,000 kilometers of navigable waterways) but remain marginal because of outdated infrastructure, lack of investment, and poor regulatory frameworks.

Lack of transportation infrastructure has hampered economic growth in the Congo Basin by creating barriers to trade not only with international markets but also internally in domestic markets. For example, domestic transport costs, at about US$3,500–US$4,500 per container, account for more than 65 percent of the total cost of importing goods to the Central African Republic (Domínguez-Torres and Foster 2011). This has created multiple landlocked economies within a single country, with limited to no exchanges among them. Deficient infrastructure holds back extractive sectors (such as logging and mining) and sectors that rely on mobility of people and goods. The agriculture sector is particularly affected, with a severe connection gap between producers from rural areas and consumers in growing urban centers.

Lack of connectivity prevents the modernization of local farming practices, with farmers unable to rely on markets for either inputs or outputs and forced to rely on self-subsistence. In the Democratic Republic of Congo, it is estimated that only 33 percent (7.6 million of the 22.5 million hectares) of all nonforested suitable arable land is less than six hours from a major market; that proportion is as low as 16 percent in the Central African Republic (Deininger et al. 2011). By contrast, 75 percent of the nonforested suitable land in Latin America is within six hours of a market town. As a result, growing domestic markets are mostly fed by food commodity imports, which deteriorate the national agriculture trade balance. Together with poor governance and high political risks, this lack of infrastructure helps explain why the Congo Basin has not seen the type of large-scale land acquisitions witnessed in other parts of the developing world. The isolation created by poor infrastructure also represents a significant risk in terms of people's vulnerability to climatic shocks. Even a modestly unsatisfactory growing season can jeopardize food security, because people have no way to benefit from surpluses in other parts of the country.

The infrastructure gap in the Congo Basin is widely acknowledged. Most Congo Basin countries have set ambitious infrastructure goals to drive economic growth and development (box O.5).

Transport infrastructure is one of the most robust predictors of tropical deforestation. Of all the different scenarios tested by the CongoBIOM, the impact of a scenario modeling improved transportation infrastructure is by far the most damaging to forest cover. The model shows that the total deforested area is three

---

Box O.5 Road Work Ahead

In response to the urgent need to upgrade their infrastructure, many countries have increased their national budget allocation to the transport sector. In the Republic of Congo, where the transportation system is by far the most deteriorated, public financing to the transport sector has increased by 31.5 percent between 2006 and 2010, with an allocation of 19.6 percent of public resources.

Significant progress has also been made to mobilize external funding to support the reconstruction of the road network. The Democratic Republic of Congo, for instance, has secured major financial commitments from multilateral and bilateral sources, including China. At the regional level, various entities are drafting plans and strategies to fill the infrastructure gap, including the Program of Infrastructure Development in Africa from the African Union/New Partnership for Africa's Development, the Consensual Road Network from Economic Community of Central African States (ECCAS), and the River Transportation plan from CICOS (*Commission Internationale du Bassin Congo-Oubangui-Sangha*). These plans define priority investments based on the development of corridors and growth poles.

Box O.6 Simulating Changes Resulting from Improved Infrastructure

The CongoBIOM was used to compute the likely impact of all the road and railway projects for which financing has already been secured. It simulated changes in average travel time to the closest city, along with changes in internal transportation costs, and took into account population density and urbanization trends. While the direct impact of road construction in rainforests is often limited, indirect and induced impacts could represent a major threat by significantly changing economic dynamics—particularly in the agriculture sector—in newly accessible areas.

A reduction in transport costs can lead to significant changes in the equilibrium of rural areas along the following causal chain:

Improved infrastructure → Increase in agriculture production → Increased pressure on forests

The model showed that when agricultural products reach urban markets at a lower price because of lower transportation costs, consumers tend to buy more domestically grown products through import substitution. This in turn encourages producers to increase their production. Additionally, the price of inputs such as fertilizers tends to go down, increasing agricultural productivity. A new equilibrium is reached with a larger volume of regionally grown agricultural products and lower prices compared with the initial situation—a change that presumably improves food security and human well-being but creates incentives for clearing forest land for agriculture. The reduction of domestic transportation costs also improves the international competitiveness of agricultural and forestry products—including products derived from uncontrolled logging along the newly opened roads.

---

times higher than in the business as usual scenario (and total greenhouse gas emissions more than four times higher because most of the deforestation would occur in dense forest). Most of the impacts do not result from the infrastructure development itself but from indirect impacts associated with higher connectivity (box O.6).

The Congo Basin's inadequate transportation infrastructure has by and large protected its forests. The challenge now is to strike the appropriate balance between forest protection and the development of a rural road network that would unlock the Congo Basin's economic potential (particularly in the agriculture sector).

Logging

Industrial logging represents an extensive land use in the Congo Basin with about 44 million hectares of forest under concession—a fourth of the total dense lowland forest area (see figure O.1). The formal logging sector produces an average of 8 million cubic meters of timber annually, with Gabon as the largest producer. Logging also contributes about 6 percent of the GDP in Cameroon, the Central African Republic, and the Republic of Congo, and is an important source of

Deforestation Trends in the Congo Basin • http://dx.doi.org/10.1596/978-0-8213-9742-8

employment in rural areas. The formal sector accounts for about 50,000 full-time jobs and much more indirect employment in the six countries. Employment created by private sector operators in the formal forestry sector is particularly important in Gabon and the Central African Republic where timber is the largest source of jobs after the public sector.

Contrary to the popular impression, logging is not uniformly a cause of deforestation and forest degradation. Ecosystem services and other land uses can coexist with logging concessions. Unlike in other tropical regions, logging in the Congo Basin usually does not result in conversion to other land uses such as cattle ranching or plantations. Industrial logging's impacts are further limited by the adoption of SFM principles as well as the high selectivity of logged species. The trend toward SFM has been momentous: As of 2010, 25.6 million hectares were managed under state-approved plans. Wood extraction rates are very low: on average, less than 0.5 cubic meter per hectare. Of the more than 100 species generally available, fewer than 13 are usually harvested.

There are significant opportunities to improve the competitiveness of the formal logging sector, so that it becomes a greater source of employment and growth. Despite the high value of their timber and gains in SFM, Congo Basin countries remain relatively small players in terms of timber production at the international level: Timber from Central Africa represents less than 3 percent of the global production of tropical roundwood, far behind the other two major tropical forest regions (OFAC 2011). Their contribution to the trade of processed timber is even smaller. Processing capacities are essentially limited to primary processing (sawn wood, peeling, and slicing for the production of plywood and veneer). Investing in modernized processing capacities along the secondary and tertiary stages could generate more added value and employment from existing forest resources and tap regional demand for higher end furniture.

Although the footprint of formal logging operations is considered low, the informal artisanal sector presents a different story. The informal sector supplies markets that are less selective than export markets; chainsaw operators are less efficient in their use of trees to produce timber; and informal activities tend to overlog the most accessible areas, surpassing regeneration rates. On the plus side, the informal sector is a larger source of direct and indirect local employment than the formal sector, with benefits more equally redistributed at the local level. Left unregulated, this segment of the forest sector may severely undermine forest biomass and erode carbon stocks.

Domestic demand for construction timber is booming and is currently quasi-exclusively supplied by the unregulated, underperforming, and unsustainable informal sector. The artisanal sector, while long overlooked, is now recognized as a major segment of the logging sector. There are few reliable data about informal logging, which is mostly oriented to domestic markets, but experts believe that it is at least as large as the formal sector. and has more serious impacts on forest loss by progressively degrading forests close to highly populated areas. In Cameroon and the Democratic Republic of Congo, informal timber production already surpasses formal timber production, and in the Republic of Congo, domestic timber production represents more than 30 percent of total timber production (Lescuyer et al. 2012). This trend is unlikely to wane as most Congo Basin countries experience a strong urbanization process. In addition, demand for informal timber emanates from other African countries (such as Niger, Chad, Sudan, the Arab Republic of Egypt, Libya, and Algeria) where demographic growth and urbanization are booming.

Mining

The Congo Basin is home to mineral resources worth billions of dollars on world markets, but that wealth has been largely untapped so far. Among these resources are valuable metals (copper, cobalt, tin, uranium, iron, titanium, coltan, niobium, and manganese) and nonmetals (precious stones, phosphates, and coal). With the exception of the Democratic Republic of Congo, the mineral wealth of the Congo Basin has been underexploited in part because of civil unrest and conflict over the past two decades, lack of infrastructure, poor business climate, and heavy reliance on oil by some countries in the region. Armed groups have often used mineral wealth to fund their activities, perpetuating a cycle of instability that discourages investment.

World demand for mineral resources increased significantly after 2000, driven by global economic development and particularly China's strong growth. While the world recession of 2008 affected the mining sector, economic recovery in some emerging countries led to a rapid revival of demand for raw materials in 2009. Growth in the technology, transportation, and construction sectors will likely continue fueling greater demand for aluminum, cobalt, copper, iron ore, lead, manganese, platinum metals, and titanium in the future.

In the context of rising demand and high prices, mineral reserves that used to be considered financially unviable are now receiving much attention. Heightened interest from investors is directly reflected in increased exploration activities in the Congo Basin including the densely forested areas. Historically, the majority of mining operations in the Congo Basin has occurred in nonforested areas, but that is projected to change. The past few years have also seen the emergence of new types of deals in which investors have offered to build associated infrastructure (including roads, railways, power plants, ports, and so on) in exchange for security of supply. The burden of the infrastructure investments is thus taken off the countries' shoulders, which theoretically alleviates one of the major constraints to mining development. At the same time, the decline of oil reserves is pushing countries like Gabon and Cameroon to develop other extractive industries to offset the revenue gap from declining oil wealth.

The mining sector could become an engine of growth in the Congo Basin. At its height in the mid-1980s, the mining sector contributed 8–12 percent of the Democratic Republic of Congo's GDP. Given the Democratic Republic of Congo's extensive copper, cobalt, gold, diamond cassiterite, and coltan reserves, mining could contribute to significant revenue increases and sustain growth in the economy as a whole, including through employment.

Deforestation Trends in the Congo Basin • http://dx.doi.org/10.1596/978-0-8213-9742-8

Box O.7 Small-Scale and Artisanal Mining and Adverse Impacts on Environment

Both artisanal miners (who operate with little mechanized aid) and small-scale miners (who use more organized and more productive methods but produce less than a certain tonnage of minerals per year) have responded to international demand for minerals by increasing activity in the Congo Basin in recent years. Some of the environmental concerns associated with artisanal and small-scale mining stem from practices that can include primary forest clearance, dam construction, the digging of deep pits without backfilling, and resulting impacts on water levels and watercourses. Forest degradation is also associated with the arrival of large numbers of migrant diggers spread out over a large area of forest. In Gabon, for example, artisanal miners suffer from a fragile legal status that gives them little incentive to pursue environmentally responsible mining (WWF 2012).

Strategies to respond to these issues include the setting up of socially responsive and environmentally sustainable supply chains and measures to professionalize and formalize artisanal and small-scale mining activities, so that risks are managed and minimum standards are introduced. These initiatives are partially inspired by the success of a third-party certification scheme called "Green Gold—Oro Verde" introduced in 1999 in Colombia to stop the social and environmental devastation caused by poor mining practices in the lush Chocó Bioregion and to supply select jewelers with traceable, sustainable metals.

The nature of the potential impacts of mining operations on forest is varied. Compared with other economic activities, mining has a fairly limited *direct* impact on forest cover. *Indirect* impacts can be more important and are tied to the larger infrastructure developments that tend to occur in a mining region, such as building power plants (including dams) and more roads. *Induced* impacts may include impacts associated with a large influx of workers, such as subsistence agriculture, logging, poaching, and other activities. Finally, *cumulative* impacts relate more to artisanal mining where many small individual mining sites add up to significant impacts (see box O.7).

Poor land use management can potentially amplify the adverse impacts of mining operations (both exploration and exploitation.) Numerous conflicts have been noted between and among conservation priorities, mining and logging concessions, and the livelihoods of local populations. For example, in the Sangha Tri-National Park (shared by Cameroon, the Central African Republic, and the Republic of Congo), projected logging and mining concessions overlap with the region's protected areas and agroforestry zones (Chupezi et al. 2009).

How to Reconcile Growth and Forest Protection Policy Options and Recommendations?

The countries of the Congo Basin face the dual challenge of urgently developing their economies to reduce poverty while limiting the negative impact on the region's natural resources. Growing international recognition of the importance of forests to stem climate change provides new opportunities for Congo Basin countries to reconcile these objectives, by leveraging climate finance and creating momentum for policy change.

Since 2007, parties to the UN Framework Convention on Climate Change (UNFCCC) have deliberated on a framework that would create incentives for reducing emissions from deforestation and forest degradation (REDD+) by rewarding tropical countries that preserve and/or enhance the carbon locked in forests. International, regional, and national discussions on the future REDD+ mechanism have given rise to a better understanding of the multiple drivers of deforestation and a more holistic view of low-carbon development in which different sectors play a role. While many elements of REDD+ remain unknown (box O.8), countries can focus on "no-regrets" measures that should yield benefits regardless of the shape of a future mechanism under the UNFCCC.

Congo Basin Countries Have the Opportunity to Embark on Development Pathways that "Leapfrog" Severe Deforestation

In December 2008, countries agreed that REDD+ reference levels should "take into account historic data and adjust for national circumstances." This appears to suggest that countries, such as those in the Congo Basin, with low historic rates of deforestation—but potentially high future rates—could consider factoring this into a proposed reference level. But credible data that would justify adjustments to historical trends could be difficult to obtain. Although the modeling approach used in this study was an attempt to use existing, limited data to offer an initial description of future deforestation trends, it was not designed to provide robust quantitative information for setting reference levels in a financing mechanism such as REDD+.

Deforestation Trends in the Congo Basin: Reconciling Economic Growth and Forest Protection highlights options to limit deforestation while pursuing

---

Box O.8 A Fair Baseline

International negotiations on forests and climate change have been positive for Congo Basin countries. The Congo Basin contains an estimated 25 percent of the total carbon stored in tropical forests worldwide and has attracted wide attention, as a result. Congo Basin countries have received support from a variety of bilateral and multilateral funds including the Forest Carbon Partnership Facility, UN-REDD, Global Environment Fund, and the Forest Investment Program. For now, financing resources fall under phase 1 of the REDD+ mechanism, which deals with the "readiness" process (including capacity building and planning). The core provision of finance is expected to come later on, in a phase that rewards measured, reported, and verified results. This could be particularly tricky in the Congo Basin context.

One of the most important challenges for Congo Basin countries relates to the development of "reference levels," or the baselines against which their success in reducing emissions will be measured. For high forest cover and low deforestation (HFLD) countries, using historic baselines may not capture the effort and economic sacrifice required to combat future deforestation risks.

---

economic growth in an inclusive and sustainable way. It outlines both cross-cutting and sector-specific recommendations. These recommendations are intended as general guidelines that should spur more detailed policy discussions at the country level.

Cross-cutting Recommendations
Invest in Participatory Land Use Planning
Participatory land use planning should be used to maximize economic and environmental objectives and reduce problems resulting from overlapping usage titles and potentially conflicting land uses. Trade-offs among different sectors and within sectors need to be clearly understood by the stakeholders so that they can define development strategies at the national level. This requires robust socioeconomic analysis as well as strong coordination among ministries and some form of high level arbitrage. Once completed, this land plan would determine the forest areas that need to be preserved, the areas that can coexist with other land uses, and those which could potentially be converted into other uses.

While planning for economic development, particular attention should be given to protect high-value forests in terms of biodiversity, watershed, and cultural values. Optimal mining, agriculture, and other activities should be directed away from forests of great ecological value. In particular, agriculture development should primarily target degraded lands. The Global Partnership on Forest Landscape Restoration estimates that more than 400 million hectares of degraded land in Sub-Saharan Africa offer opportunities for restoring or enhancing the functionality of "mosaic" landscapes that mix forest, agriculture, and other land uses.

One output of land planning could be the identification of growth poles and major development corridors that could be developed in a coordinated manner, with the involvement of all government entities along with the private sector and civil society. In the Congo Basin, this approach would likely be driven by natural resources and provide upstream and downstream linkages around extractive industries. While a land use planning exercise definitely needs to be conducted at the country level (and even at the provincial level), the corridor approach has also been adopted by the Economic Community of Central African States (ECCAS) at the regional level to foster synergies and economies of scale among member states.

Improve Land Tenure Schemes
Effective systems of land use, access rights, and property rights are essential to improve the management of natural resources. Improving these systems is a priority for providing farmers, especially women, with the incentives needed to make long-term investments in agricultural transformation. Likewise, there is strong evidence that community-based forest management approaches can expand the supply of woodfuel and relieve natural forests from unsustainable withdrawals, provided communities are given enough visibility on land/tree

tenure issues to invest in the long-term sustainability of forests, woodlots, and agroforestry systems.

Current land tenure schemes are not conducive to grassroots SFM in Congo Basin countries. Outside of commercial logging concessions, forests are considered "free access" areas under state ownership and are not tagged with property rights. Moreover, tenure laws in most Congo Basin countries directly link forest clearing (*mise en valeur*) with land property recognition and thus create an incentive to convert forested lands into farmland. Current land tenure laws should be adjusted to separate land property recognition from forest clearing.

Strengthen Institutions

Without strong institutions able to enforce rules and build alliances within a complex political economy, neither land use planning nor tenure reform will yield real change. Administrations face expectations—in terms of planning, monitoring, and controlling forest resources—that they cannot adequately meet when they are weak. Properly staffed and equipped institutions are necessary to fight illegal activities but also to undertake the difficult tasks of formalizing artisanal logging, the woodfuel/charcoal value chain, and artisanal mining in critical ecosystems. New technologies (based on geographic and information technology systems) should be more widely available to administrations to improve their performance.

To succeed, REDD+ needs to build on strong institutions, notably in terms of law enforcement and monitoring. To get ready for phase 3 financing of REDD+ institutions, the Congo Basin countries will have to be able to set up credible monitoring systems so the international community can track progress made in specific countries. Monitoring efforts are to be performed by regulatory agencies, but strategic partnerships can be set up to improve monitoring activities: Local communities can be trained and engaged in helping regulators monitor activities on the ground; nongovernmental organizations (NGOs) can also provide additional monitoring via field projects, for example, near mining sites.

REDD+ should built and strengthen existing processes such as the Comprehensive Africa Agriculture Development Plan (CAADP) and the Forest Law Enforcement, Governance, and Trade (FLEGT) initiatives. CAADP provides an excellent and timely opportunity to thoroughly analyze agricultural potential, develop or update national and regional agricultural investment plans aimed at increasing agricultural productivity on a sustainable basis, and strengthen agricultural policies. For the forest sector, the FLEGT approach, backed by the European Union in all Congo Basin countries except Equatorial Guinea, provides an effective vehicle for improving forest governance, including the domestic arena.

Recommendations by Sector
Agriculture: Increase Productivity and Prioritize Nonforested Lands
- Prioritize agricultural expansion on nonforested areas. There is an estimated 40 million hectares of suitable noncropped, nonforested, nonprotected land

in the Congo Basin. This corresponds to more than 1.6 times the area currently under cultivation. Utilizing these available areas, along with an increase in land productivity, could dramatically transform agriculture in the Congo Basin without taking a toll on forests. Decision makers must prioritize expanding agriculture on nonforested lands.

- Empower smallholder farmers. With about half the population active in agriculture in most countries of the Congo Basin, there is a need to foster sustained agricultural growth based on smallholder involvement. Experience in other tropical regions shows this is possible. Thailand, for example, considerably expanded its rice production area and became a major exporter of other commodities by engaging its smallholders through a massive land-titling program and government support for research, extension, credit, producer organizations, and rail and road infrastructure development.

- Reinvigorate research and development (R&D) toward sustainable productivity increase. R&D capacities in the Congo Basin, with the exception of Cameroon, have been dismantled over the past decades. Research has largely neglected the most common staple food crops in the Congo Basin, such as yams, plantains, and cassava, usually referred to as neglected crops. The potential to increase productivity of these crops and improve their resistance to disease and tolerance to climatic events has also been untapped. Partnerships need to be established with international research centers (for example, among members of the Consultative Group on International Agriculture Research) to stimulate agricultural research in the Congo Basin and progressively strengthen national capacities.

- Promote a sustainable large-scale agrobusiness industry. Large agribusiness operations, especially rubber, oil palm, and sugarcane plantations, have the potential to sustain economic growth and generate considerable employment for rural populations. Given weak land governance, there is a risk that investors will acquire land almost for free, interfere with local rights, and neglect their social and environmental responsibilities. Governments should establish stronger policies on future large land investments, including requiring land applications to be oriented toward abandoned plantations and suitable nonforested land. Efforts to put palm oil production on a more sustainable footing, such as the Roundtable on Sustainable Palm Oil founded in 2004, may help mitigate some of these environmental issues by setting standards that aim to prevent further loss to primary forests or high conservation value areas and reduce impacts on biodiversity.

- Foster win–win partnerships between large-scale operators and smallholders. Such partnerships could become an engine for transforming the agricultural sector. While this has not yet materialized in the Congo Basin, there are many examples in the world where meaningful partnerships between smallholders and large-scale operators have yielded successful results and promoted a well-balanced development of agriculture.

Energy: Organize the Informal Value Chain

- Put woodfuel energy higher on the political agenda. Despite its undisputed importance as the major source of energy, woodfuel is still getting very little attention in the policy dialogue and therefore is poorly featured in the official energy policies and strategies. There is a need to change the policy makers' perception of wood energy as "traditional" and "old-fashioned." Lessons could be drawn from Europe and North Africa where wood energy is emerging as a cutting edge renewable energy source. Congo Basin countries should seize technical breakthroughs and climate finance opportunities to put this energy resource on a more modern and efficient footing.

- Optimize the woodfuel/charcoal value chain. Formalizing the sector would break the oligopolistic structure of the sector and create a more transparent marketplace. The economic value of the resources would thus be better reflected in the pricing structure and appropriate incentives could be established. Such formalization should be supported by the revision and modernization of the regulatory framework. To do so, the Congo Basin countries would have to understand the political economy of the woodfuel/charcoal value chain. A multistakeholder dialogue would be critical to help overcome difficult trade-offs between sustaining rural livelihoods based on informal activities and enforcing production standards and trade restrictions that would come with the formalization of the sector.

- Diversify supplies. The charcoal value chain in the Congo Basin currently relies exclusively on natural forests. Although natural forests are expected to continue supplying much of the raw material for charcoal production, they will be unable to meet growing demand in a sustainable manner. Policy makers should consider diversifying the sources of wood, by increasing sustainable wood supply through tree plantations and agroforestry and maximizing the potential supply from natural forests, with special attention to timber waste management.

- Foster community involvement through devolution of rights and capacity building. Community-based woodfuel production schemes in Niger, Senegal, Rwanda, and Madagascar have shown promising results when long-term rights to forest land and devolution of management have motivated communities to participate in woodfuel production. Pilots have been launched in the Congo Basin (cf. Batéké plantations) and could be replicated.

- Respond to growing urban needs in terms of both food and energy. Deforestation and forest degradation mostly occur around urban centers in the Congo Basin countries, as a result of ad hoc agricultural expansion to respond to rising demand for food and energy. An integrated, multiuse approach to meeting urban needs would address the various driving forces of forest degradation. If well organized, it could not only secure the food and energy needs for a growing urban population but could also provide sustainable solutions to unemployment and waste management.

Transportation: Better Plan to Minimize Adverse Impacts

- Improve transportation planning at local, national, and regional levels.

  Locally: Areas that are directly served by improved transportation facilities will become more competitive for various economic activities such as agricultural expansion including palm oil plantations. Local participation in transportation planning will help ensure that economic opportunities are maximized. Mitigation measures at the local level may include clarifying land tenure or integrating the transportation project into a broader local development plan. Such plans may include the protection of forest banks along roads, rivers, or railways to avoid unplanned deforestation. Defined up front and in a participatory manner, these restrictions would get more backing from the different stakeholders.

  Nationally and regionally: The corridor approach shows that improving transportation services (for example, freight management in harbors) or infrastructure (facilitating river or rail transportation) may have a wider macroeconomic impact at the regional level. Planning at the national and regional levels through a corridor approach could help identify adequate mitigation measures, such as zoning reforms (establishing permanent forest areas), law enforcement (ensuring the respect of zoning decisions), land tenure clarification, and controlling the expansion of agriculture

- Foster a multimodal transport network. As countries plan for transport development, it is important that they consider the pros and cons of roads and alternative transport modes such as navigable waterways and railroads, not only in terms of economic returns but also in terms of environmental impacts. For instance, with more than 12,000 kilometers of navigable network, the Congo Basin could benefit from a potentially highly competitive waterway system.

- Properly assess ex ante impacts of transport investments. Transportation development (be it new infrastructure or rehabilitation of existing assets) will reshape the economic profile of the areas served by transportation and increase pressure on forest resources. Currently, most environmental impact studies or safeguard reviews fail to fully capture the long-term indirect effects on deforestation. New assessment methods, based on economic prospective analysis, could help prioritize infrastructure investments with low foreseen impacts on forests.

Logging: Expand SFM to the Informal Sector

- Pursue progress on SFM in industrial logging concessions. Although the Congo Basin already has vast concession areas under management plans, further progress can be made through ensuring adequate implementation; adjusting SFM standards and criteria to reflect climate change and advances in reduced impact logging techniques; moving away from single-use, timber-oriented management models; encouraging certification schemes; and supporting the FLEGT process.

- Foster the involvement of communities in forest management. Although the concept of "community forestry" has been embraced by most Congo Basin countries and entered their legal framework, shortfalls such as time-bound management contracts continue to constrain effective community forest management of state-owned forests. Revisiting the concept and clarifying community rights over forests could provide a window of opportunity to revitalize its implementation on the ground.

- Formalize the informal timber sector. To ensure a sustainable supply of timber for domestic markets and spread SFM principles to the domestic timber market, numerous small and medium forest enterprises will need to be supported by appropriate regulations. For the woodfuel/charcoal value chain, such formalization would rely on an in-depth understanding of the political economy of the sector and would require an open dialogue with various stakeholders. In addition, domestic and regional timber markets will have to be better understood to help decision makers respond to market opportunities without jeopardizing natural forest assets.

- Modernize processing capacities to set up an efficient timber value chain in the Congo Basin with less waste and more domestic value added. The development of the secondary and tertiary processing industry would allow Congo Basin countries to use secondary tree species to respond to domestic growing needs.

Mining: Set "High-Standard" Goals for Environmental Management

- Properly assess and monitor impacts of mining activities. Proper environmental impact assessments and social impact assessments have to be prepared for all stages of mining operations (from exploration to mine closure); management plans also need to be of a good quality and their implementation regularly monitored to mitigate the associated risks.

- Learn from international best practices and foster risk mitigation. If mining is to result in minimal negative impacts to the forests of the Congo Basin, companies will need to follow best international practices and standards designed to meet the mitigation hierarchy (Avoid—Minimize—Restore—Compensate). International standards for responsible mining have been developed by various organizations, including the International Council on Mining and Metals, the Responsible Jewelry Council, the International Finance Corporation, and the Initiative for Responsible Mining Assurance. Lessons can be learned from these innovative approaches as governments adjust their national regulations on mining activities and their environmental monitoring and management.

- Upgrade the artisanal and small-scale mining sector. Efforts should focus on putting small-scale miners on a more secure footing and adjusting regulatory frameworks so that they can better respond to the specific needs of this segment of the mining sector. Governments should facilitate the use of environmentally friendly technologies and encourage the development of a sustainable supply chain. The Alliance for Responsible Mining has developed a certification system for small-scale mining cooperatives that includes consideration of both environmental and social concerns. The Green Gold approach is another example.

- Promote innovative mechanisms to offset negative impacts of mining operations. Conservation groups have advocated for biodiversity offsets for extractive projects for at least a decade. Financial instruments, such as financial guarantees, could also be options to mitigate adverse impacts, particularly to ensure mine reclamation and restoration at the closure of the mining site.

Note

1. From 1,094,000 to 1,301,000 metric tons according to the UN Energy Statistics Database.

References

Chupezi, T. J., V. Ingram, and J. Schure. 2009. *Study on Artisanal Gold and Diamond Mining on Livelihoods and the Environment in the Sangha Tri-National Park Landscape, Congo Basin.* Yaounde, Cameroon: Center for International Forestry Research/ International Union for Conservation of Nature.

Deininger, K., and D. Byerlee, with J. Lindsay, A. Norton, H. Selod, and M. Stickler. 2011. *Rising Global Interest in Farmland: Can It Yield Sustainable and Equitable Benefits?* Washington, DC: World Bank.

de Wasseige C., P. de Marcken, N. Bayol, F. Hiol Hiol, P. Mayaux, B. Desclée, R. Nasi, A. Billand, P. Defourny, and R. Eba'a Atyi. 2012. *The Forests of the Congo Basin—State of the Forest 2010.* Luxembourg: Publications Office of the European Union.

Dominguez-Torres, C., and V. Foster. 2011. *The Central African Republic's Infrastructure: A Continental Perspective.* Washington, DC: World Bank.

Hansen, M., S. Stehman, P. Potapov, T. Loveland, J. Townshend, R. Defries, K. Pittman, B. Arunarwati, F. Stolle, M. Steininger, M. Carroll, and C. DiMiceli. 2008. "Humid Tropical Forest Clearing from 2000 to 2005 Quantified by using Multitemporal and

Multiresolution Remotely Sensed Data." *Proceedings of the National Academy of Sciences of the United States of America* 105 (27): 9439–44.

Hoyle, D., and P. Levang. 2012. "Oil Palm Development in Cameroon." Ad Hoc Working Paper, World Wildlife Federation in partnership with Institut de Recherche pour le Développement and Center for International Forestry Research.

IEA (International Energy Agency). 2006. "Prospectives énergétiques mondiales 2006. WEO." Organisation pour le Coopération économique et le Développement OCDE/AIE Paris, France.

IFPRI (International Food Policy Research Institute). 2011. *2011 Global Hunger Index*. Washington, DC: IFPRI. Available at http://www.ifpri.org/publication/2011-global-hunger-index.

Lescuyer, G., P. O. Cerutti, E. Essiane Mendoula, R. Eba'a Atyi, and R. Nasi. 2012. "An Appraisal of Chainsaw Milling in the Congo Basin." In *The Forests of the Congo Basin—State of the Forest 2010*, ed. by C. de Wasseige et al. Luxembourg: Publications Office of the European Union.

Marien, J-N. 2009. "Peri-Urban Forests and Wood Energy: What Are the Perspectives for Central Africa?" In *The Forests of the Congo Basin—State of the Forest 2008*, ed. by C. de Wasseige et al. Luxembourg: Publications Office of the European Union.

OFAC (Observatory for the Forests of Central Africa). 2011. *National Indicators*. Kinshasa: OFAC. Available at http://www.observatoire-comifac.net (accessed December 2011).

Peltier R, F. Bisiaux, E. Dubiez, J-N. Marien, J-C. Muliele, P. Proces, and C. Vermeulen. 2010. "De la culture itinérante sur brûlis aux jachères enrichies productrices de charbon de bois, en Rep. Dem. Congo." In *Innovation and Sustainable Development in Agriculture and Food 2010 (ISDA 2010)*. Montpellier, France.

UNDP (United Nations Development Programme). 2012. *The Africa Human Development Report 2012: Building a Food Secure Future*. New York: UNDP.

WWF (World Wildlife Fund). 2012. *Gabon Case Study Report*. Artisanal and Small-Scale Mining in and around Protected Areas and Critical Ecosystems Project (ASM-PACE). Washington, DC: WWF.

Introduction

The Congo Basin represents 70 percent of the African continent's forest cover and constitutes a large portion of Africa's biodiversity. The nations of Cameroon, the Central African Republic, the Democratic Republic of Congo, Republic of Congo, Equatorial Guinea, and the Gabon share the Basin's ecosystem. About 57 percent of the Basin is covered by forest. It is the second largest tropical forest area in the world, behind only the Amazon forests.[1] The Congo Basin forest performs valuable ecological services, such as flood control and climate regulation, at the local and regional levels. The wealth of carbon stored in the Basin's abundant vegetation further serves as a buffer against global climate change. Congo Basin forests are home to about 30 million people and support livelihoods for more than 75 million people from more than 150 ethnic groups who rely on local natural resources for food, nutritional, health, and livelihood needs. In all of the six countries, forestry is a major economic sector, providing jobs and local subsistence from timber and nontimber products and contributing significantly to export and fiscal revenues.

Historically, the Congo Basin forest has been under comparatively little pressure, but there are signs that this situation is likely to change as pressure on the forest and other ecosystems increases. Until very recently, low population density, unrest and war, and low levels of development hampered conversion of forests into other land uses; however, satellite-based monitoring data now show that the annual rates of gross deforestation in the Basin have doubled since 1990. There is indeed some evidence that the Basin forests may be at a turning point of heading to higher deforestation and forest degradation rates. The forest ecosystems have not yet suffered the damage observed in other tropical regions (Amazonia, Southeast Asia) and are quite well preserved. The low deforestation rates mainly result from a combination of such factors as poor infrastructure, low population densities, and political instability that have led to the so-called passive protection. However, signs that the Congo Basin forests could be under increasing pressure from a variety of forces—both internal and external—range from mineral extraction, road development, agribusiness, and biofuels to agriculture expansion for subsistence and population growth. All of these factors could drastically amplify the pressure on natural forests in the coming decades and trigger the transition from the "high forest/low deforestation" profile into a more intense pace of deforestation.

The growing recognition of the contribution of forests to climate change has created new momentum in the fight against tropical deforestation and forest degradation. The inclusion of forests within international agreements on climate change, particularly under the United Nations Framework Convention on Climate Change (UNFCCC), coupled with commitments from developed countries to provide technology, capacity building, and finances to help developing countries tackle climate change, presents a new opportunity for forested developing countries. Since 2005, parties to the UNFCCC have deliberated on a framework that would create incentives for "reducing emissions from deforestation, forest degradation, forest conservation, the sustainable management of forests, and enhancement of forest carbon stocks" (REDD+). The conference of the parties has adopted various relevant decisions to establish basic parameters and rules for the creation of REDD+ accounting frameworks and implementation guidelines.

In order to be successful, REDD+ has to be anchored in the context of sustainable and low-emission development strategies. A future REDD+ mechanism should be central in helping countries identify new ways to reconcile their urgent need to transform their economies and the preservation of their forests, which is considered a Global Public Good. A REDD+ framework embedded in a broader economic growth strategy can create important incentives to protect natural resources in the Congo Basin while also promoting its sustainable development and tackling the key drivers of deforestation—the majority of which are outside the forest sector. Most Congo Basin countries are actively engaged in the process of developing REDD+ frameworks, or strategies, and are already working to improve their capacity to monitor forest-related emissions, improve forest governance, promote development, and reduce poverty while protecting the region's natural resources. This is being done with multilateral programs, such as the Forest Carbon Partnership Facility (FCPF) and the UN-REDD+ Program, and bilateral programs.

The objective of the two-year exercise was to analyze and get a better grasp of the deforestation dynamics in the Basin. The primary goal of the exercise was to give stakeholders (and particularly policy makers) a thorough understanding of how economic activities (agriculture, transport, mining, energy, and logging) could impact the region's forest cover through an in-depth analysis of the connections between economic developments and forest loss. The exercise was not meant, however, to achieve quantified deforestation predictions. The approach used for this analysis relied on a combination of robust analytical work, a modeling exercise, and regular and iterative consultation with technical experts from the region. The modeling tool was chiefly used to better comprehend the chain of causes and effects between various economic forces and their potential impacts on forest cover and, thus, carbon content. This exercise yielded significant progress toward understanding the various drivers of deforestation and was particularly useful in better reflecting the impact of economic forces exogenous to the Congo Basin.

The report combines the outcomes of the modeling exercise with in-depth sector analyses on agriculture, transport, energy, mining, and logging. It also reflects the results of a modeling exercise conducted by the International Institute for Applied Systems Analysis (IIASA) to better understand the national, regional, and international drivers of deforestation. The report was generated through a highly interactive process with stakeholders from the Congo Basin to identify specific needs of the relevant countries. Accordingly, the partial equilibrium model elaborated by IIASA (GLOBIOM model) was downscaled to the Congo region as part of this exercise; the CongoBIOM model is now available for Congo Basin countries to predict drivers and patterns of deforestation under their own data set.

The structure of this report is as follows:

- Chapter 1 gives an overview of the forests of the Congo Basin, including an analysis of the historical trends of deforestation and forest degradation.
- Chapter 2 presents the dynamics of deforestation and summarizes the results of a sector-by-sector analysis of significant drivers of deforestation in the Basin, including agriculture, logging, energy, transportation, and mining.
- Chapter 3 provides an update on the state of negotiations under the UNFCCC on REDD+ and implications for Congo Basin countries—covering, in particular, some of the key opportunities as well as the challenges for "high forest cover, low deforestation" countries. The chapter builds on the analysis of the previous chapters and recommends priority activities for Basin countries in order to address the current and future drivers of deforestation.

Note

1. For more information, see http://www.fao.org/docrep/014/i2247e/i2247e00.pdf; accessed March 8, 2012.

CHAPTER 1

Congo Basin Forests: Description

The Forest Ecosystems in the Congo Basin

The Congo Basin forest is the world's second-largest contiguous block of tropical forest. The Basin encompasses 400 million hectares, 200 million of which are covered by forest, with 90 percent being tropical dense forests. More than 99 percent of the forested area is primary forest or naturally regenerated forest, as opposed to plantations. The Congo Basin forest, also referred to as the Lower Guineo-Congolian forest, extends from the coast of the Atlantic Ocean in the west to the mountains of the Albertine Rift in the east. It spans the equator by nearly seven degrees north and south (CBFP 2005). Eighty percent of the Congo Basin forest is located between 300 and 1,000 meters altitude, and scientists have divided it into six ecological regions[1] that signal priority areas for conservation (Olson and Dinerstein 2002) (map 1.1).

Dense forests represent the largest portion of land cover, with about half (46 percent) of them classified as dense humid forests; woodlands comprise about one-fifth of the land cover. Of the remainder, about 8 percent is a forest–agriculture mosaic. Dense forests are divided into different categories (lowland forests: 900 meters or more; submontane forests: between 900 and 1,500 meters; montane forests: less than 1,500 meters; and edaphic forests and mangroves). In all Congo Basin countries except the Central African Republic, dense forests represent the most extensive land cover, from 40 percent in Cameroon to 84 percent in Gabon. Table 1.1 summarizes the area estimates of different land-cover classes (de Wasseige et al. 2012).

The distribution of these forest types correlates strongly with annual rainfall. The northern forests have a hot, severe dry season, while the rest—particularly those in the west—have much cooler dry seasons. Along the Atlantic coast, to the west, extends a belt of species-rich evergreen forest. This forest, the wettest in the region, experiences annual rainfall in excess of 3,000 millimeters in some areas and extends inland for a distance of about 200 kilometers, after which the forest becomes progressively drier and species-poor nearer the interior. The swamps ecoregions, found in the center of the forest block, support

Map 1.1 Forest Ecosystems in the Congo Basin and Their Biodiversity

Source: Authors, based on the World Wildlife Fund's (WWF) major ecosystems of the Congo River Basin 2012.
Note:
Cross-Sanaga-Bioko coastal forests: Forest elephants and many primates such as Cross River gorillas and chimpanzees, other primates (some of which are endemic to the ecoregion); great variety of amphibians (including the goliath frog), reptiles, and butterflies.
Atlantic Equatorial coastal forests: Western gorilla, elephant, mandrill, and other primates; evergreen forest diversity including plants, birds, and insects.
Northwestern Congolian lowland forests: High wildlife densities, western lowland gorilla, elephant, bongo; plant diversity low in the east and higher in the west.
Western Congolian swamp forests: Wetlands fauna and flora, elephant, western lowland gorilla, chimpanzee, and other primates; low plant diversity and some wetlands endemics.
Eastern Congolian swamp forests: Wetlands fauna and flora, bonobo, and other primates; low plant diversity and some wetlands endemics.
Central Congolian lowland forests: Bonobo, okapi, elephant, salongo monkey, and other primates; plant diversity apparently low.
Northeastern Congolian lowland forests: Grauer's gorilla, okapi, owl-faced monkey, other primates, and birds; fairly high plant diversity.

Table 1.1 Area Estimates (ha) of Land-Cover Types for the Six Congo Basin Countries

Land-cover class	Cameroon	Central African Republic	Congo, Dem. Rep.	Congo, Rep.	Equatorial Guinea	Gabon	% of total land
Lowland dense, moist forest	18,640,192	6,915,231	101,822,027	17,116,583	2,063,850	22,324,871	41.83%
Submontane forest	194,638	8,364	3,273,671	–	24,262	–	0.87%
Montane forest	28,396	–	930,863	10	6,703	19	0.24%
Edaphic forest	–	95	8,499,308	4,150,397	–	16,881	3.14%
Mangrove forest	227,818	–	181	11,190	25,245	163,626	0.11%
Total dense forest	**19,091,044**	**6,923,690**	**114,526,050**	**21,278,180**	**2,120,060**	**22,505,397**	**46.18%**
Forest–savanna mosaic	2,537,713	11,180,042	6,960,040	517,068	–	51,092	5.26%
Rural complex and young secondary forest	3,934,142	713,892	21,425,449	3,664,609	507,281	1,405,318	7.84%
Tropical dry forest– miombo	1,292,106	3,430,842	23,749,066	297,824	172	31,337	7.13%
Woodland	11,901,697	34,381,438	36,994,935	2,659,375	4,669	787,231	21.48%
Shrubland	2,561,163	4,002,258	6,705,478	2,101,556	1,308	619,347	3.96%
Grassland	177,385	62,015	4,372,677	1,191,956	86	341,688	1.52%
Others	*4,668,275*	*1,152,349*	*17,714,723*	*2,482,305*	*30,592*	*685,838*	*6.62%*
Total	**46,163,525**	**61,846,526**	**232,448,418**	**34,192,873**	**2,664,168**	**26,427,248**	**100.00%**

Source: Prepared from data in Food and Agriculture Organization of the United Nations (FAO) 2011 and de Wasseige et al. 2012.

significant plant and animal endemism in a vast mosaic of wetlands and riparian vegetation types. At the eastern edge of the Central African forests, the terrain rises toward the mountains of the Albertine Rift (CBFP 2005).

Biodiversity in the Congo Basin Forests

The Congo Basin forests harbor an extraordinary biodiversity with a very high level of endemism. The Congo's coastal forests are likely the most diverse in the Afrotropics, but information remains scarce for several ecoregions in the central Congo Basin (Billand 2012). The flora in the low-altitude forests is made up of more than 10,000 species of higher plants, of which 3,000 are endemic. Montane forests are home to 4,000 species, with at least 70 percent of them endemic. The Congo Basin forests also house African elephants and buffalo alongside such endemic species as the okapi, the bongo, the bonobo, and the gorilla as well as many endemic species of birds. The flora and fauna are, however, unequally distributed, and thus species richness varies between regions. The areas with the greatest variety of species are the forests of Lower Guinea in the west (Cameroon, Equatorial Guinea, and Gabon) and those of the Albertine Rift in the eastern part of the Democratic Republic of Congo (CBFP 2006; Ervin et al. 2010).

- The Atlantic Equatorial coastal forests as well as the montane and submontane forests along the mountains east of the Congo Basin exhibit high levels of biodiversity and are also most at risk. The Atlantic Equatorial coastal forest ecoregion is somewhat small in size and has been under human pressure for a relatively longer time. It is characterized by a significant degree of biodiversity in dense coastal forests that include evergreen and semideciduous formations at elevations less than 300 meters. The evergreen montane and submontane forests occur at altitudes higher than 1,000 meters. Trees are smaller, and stem density is greater, but the composition of species is relatively less diverse. In the Congo Basin, the main regions of montane or submontane forests are found in the Albertine Rift and in coastal Central Africa (WWF 2012).

• Most of the central Congo Basin forest—formed by the Northwestern, Northeastern, and Central Lowland ecoregions—consists of a mosaic of moist, dense evergreen and drier semi-evergreen formations, which are generally less varied in species. These forests have high canopies that block light, limit the growth of shrubs and grass, and favor epiphytes. The upper layer (35–45 meters) of evergreen forests in the central part of the Basin is dominated by a few species, such as *Gilbertiodendron dewevrei*, *Julbernadia seretii*, and *Brachystegia laurentii*. In the center of the Basin are 220,000 square kilometers of swamp forests or floodplain forests that exhibit less diversity but a fairly substantial degree of plant endemism (CBFP 2006).

• The borders of the Congo Basin are characterized by semideciduous, tropical dry (miombo) forest. At the boundaries of the Basin, deciduous trees constitute the upper layer of the forest (up to 70 percent) mixed with evergreen species. Semideciduous forest occurs in areas where dry periods last at least three months and where trees lose their leaves during that time. The forests are richer in plant species than are evergreen forest and are characterized by a mixture of species dominated by hackberry (*Celtis spp.*), samfona (*Chrysophyllum perpulchrum*), and bark cloth tree (*Antiaris welwitschii*), among others. The canopy of this kind of forest is characteristically undulating. Many commercial species are found in the semideciduous forest (for example, *Meliaceae*, *Tryplochiton scleroxylon*, *Chlorophora excelsa*) of southeastern Cameroon, the Central African Republic, and northern Republic of Congo. These forests give way to a mosaic of savannas and gallery forests, less rich from a botanical point of view but with greater populations of large mammals.

Ecological Services: From Local to Global

The Congo Basin forest performs valuable ecological services at local and regional levels. Such services include maintaining the hydrological cycle (water quantity and quality) and controlling flood in a high-rainfall region. The biodiversity of the forest provides timber, nontimber products, food, and medicine to millions. An additional regional benefit is regional-scale climate regulation, crucial for fostering resilience to climate change. Healthy forest ecosystems facilitate regional-scale cooling through evapotranspiration and provide natural buffers against regional climate variability (West et al. 2011; Chapin et al. 2008).

Table 1.2 Carbon Stocks in the Congo Basin Forests, 1990–2010

	Total carbon stock (million tons)			Annual change rate (%)	
	1990	2000	2010	1990–2000	2000–10
Carbon in biomass	37,727	36,835	35,992	−0.24	−0.23
Carbon in deadwood	3,115	2,923	2,664	−0.64	−0.92
Carbon in litter	665	648	634	−0.26	−0.22
Carbon in soil	18,300	17,873	17,452	−0.24	−0.24
Total carbon stock	59,807	58,279	56,742	−0.26	−0.27

Source: Authors, adapted from FAO 2011.

Congo Basin forests also provide ecological services to the global population through their capacity to store huge amounts of carbon. Tropical forests harbor one-fourth of the total terrestrial carbon stock found in the vegetation and soil (Houghton et al. 2001). While figures vary, the estimate for total carbon stored in the Basin is almost 60 billion metric tons, the largest portion of which is contained in the Democratic Republic of Congo (table 1.2). Biomass carbon accounts for about 63 percent of the total carbon stocks, followed by soil carbon (20 percent), deadwood (5 percent), and litter (1 percent). As far as change in carbon stocks is concerned, it has been more pronounced in deadwood over the last two decades. So far, the annual rate change in carbon stocks has been relatively modest.

Dense humid forests account for the majority (65 percent) of the total carbon stock in the Congo Basin forests. Closed evergreen lowland forests are a carbon gold mine, accounting for more than 90 percent of the total carbon stock within the dense humid forests category. Swamp forests represent less than 6 percent of the total carbon stocks in the Basin; montane and submontane forests account for only 0.4 and 2.6 percent, respectively (Nasi et al. 2009).

Forests play an important role in the cycling of greenhouse gas (GHG), acting as both a sink and a source of carbon dioxide,[2] methane, and nitrous oxide. Forest ecosystems' place in the global carbon cycle has gained more prominence with the world's concern for climate change. Forest ecosystems, particularly in the tropics, influence the global climate as major contributors to the global terrestrial carbon sink, which absorbs about 30 percent of all CO_2 emissions every year and additionally stores large reservoirs of carbon (Canadel and Raupach 2008). However, deforestation and degradation of these forests due to logging is also a source of carbon emissions (see box 1.1). Roughly between 10 and 25 percent of anthropogenic emissions worldwide result from the loss of natural forests; the entire global transportation sector emits fewer (DeFries, Houghton, and Hansen 2002; Hansen et al. 2008; and Harris et al. 2012). For the purposes of REDD+ (reducing emissions from deforestation and forest degradation; see chapter 3) and climate negotiations, deforestation and degradation are usually considered only in terms of carbon stocks; however, deforestation and degradation obviously have significant impacts on biodiversity as well as other forest functions.

Box 1.1 Variations in Forest Carbon Stocks: Key Concepts

Deforestation is the long-term or permanent conversion of forest land into other, nonforest uses. The United Nations Framework Convention on Climate Change defines deforestation as "*the direct, human-induced conversion of forested land to non-forested land.*"[a] This transformation can result from an abrupt event (deforestation = forest → nonforest), in which case the change in land cover and land use occurs immediately and simultaneously, or it can follow a process of progressive degradation (deforestation = forest → degraded forest → nonforest). Deforestation occurs when at least one of the parameter values used to define "forest land" is reduced from above the threshold for defining "forest" to below this threshold, for a period of time that is longer than the period of time used to define "temporarily unstocked."

Forest degradation is "forest land remaining forest land and continuing to meet the basic national criteria related to minimum forest area, forest height, and tree crown cover" but gradually losing carbon stocks as a consequence of direct human intervention (for example, logging, woodfuel collection, fire, grazing). "Degradation" is thus the conversion of a forest

Figure B1.1.1 Forest Degradation and Deforestation: Variation of Carbon Stocks in Above-Ground Biomass

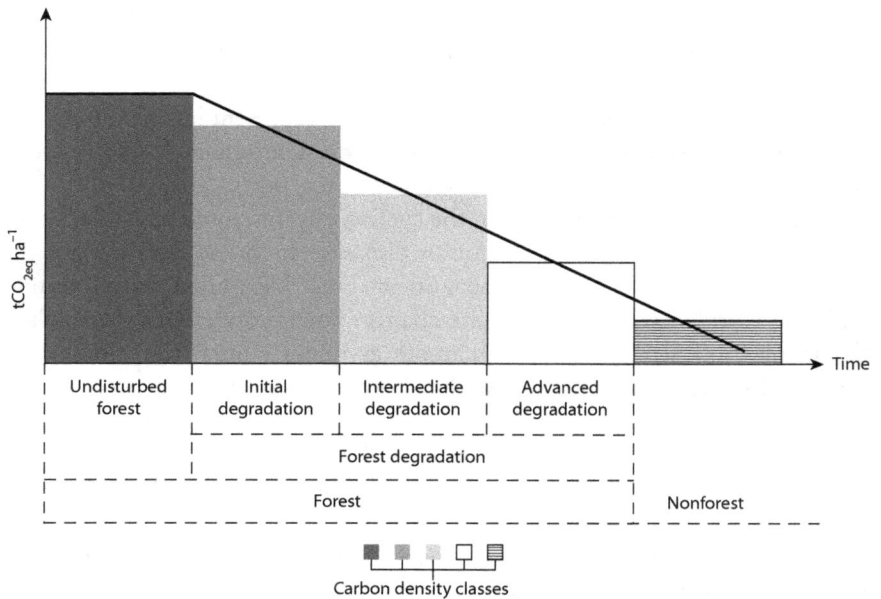

Note: tCO$_2$e ha^{-1} = tons equivalent of carbon dioxide per hectare.

Box 1.1 Variations in Forest Carbon Stocks: Key Concepts *(continued)*

class with higher average carbon stock density into one with lower average carbon stock density.

Consistent with the above definition, areas subject to **sustainable forest management** (with logging activities) represent a particular class of "degraded forest." An undisturbed natural forest that will be subject to sustainable forest management will lose part of its carbon but the loss will partially recover over time. In the long term, a sustainable harvesting and regrowth cycle could maintain a constant average carbon stock density in the forest. Since this average carbon stock density is lower than that in the original forest, sustainably managed forests are considered a special case of "degraded forest."

Figure B1.1.2 Sustainable Forest Management: Variation of the Carbon Stocks in Above-Ground Biomass

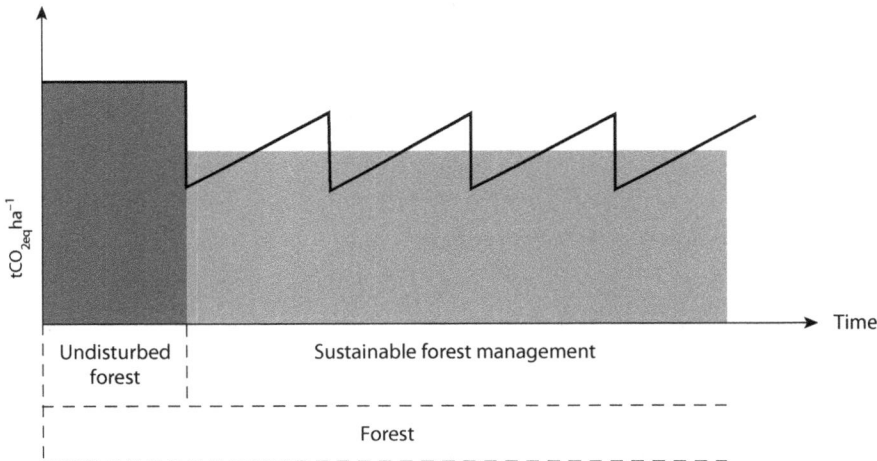

Forest regeneration corresponds to a transition from a disturbed forest class to a forest class with higher carbon stock density. Degraded forests or young forests (planted or secondary) can increase their carbon stock if properly managed or when logging and other activities are permanently suspended or reduced.[b] The process can be seen as the reversal of forest degradation.

Reforestation/afforestation is a specific case of forest regeneration when the initial status of the land is nonforest land. Depending on whether the land was a forest before or after 1990, the mechanisms of forest regeneration are called afforestation or reforestation, respectively.

box continues next page

Box 1.1 Variations in Forest Carbon Stocks: Key Concepts *(continued)*

Figure B1.1.3 Forest Regeneration: Variation of the Carbon Stocks in Above-Ground Biomass

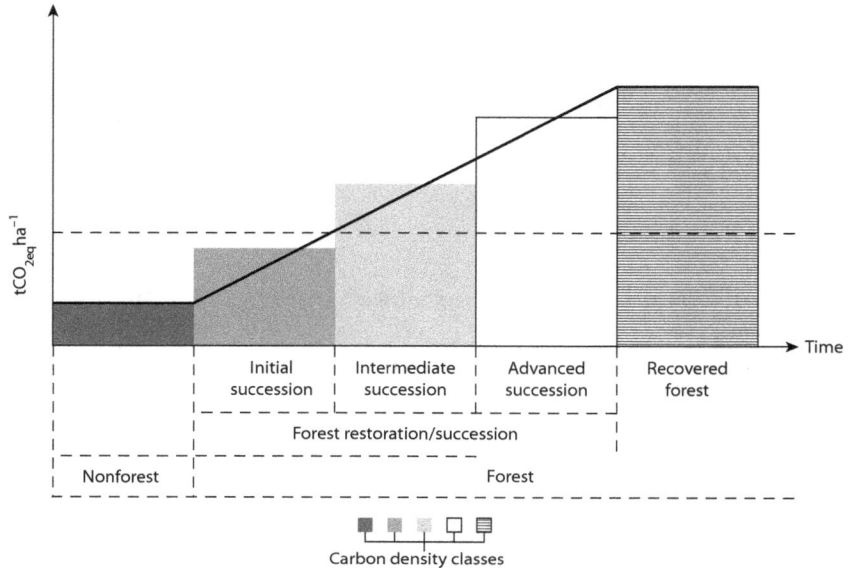

Note: tCO_2e ha^{-1} = tons equivalent of carbon dioxide per hectare.

Note:
a. Forest-area and carbon-stock losses due to natural disturbances (landslides, volcanic eruptions, and sea-level rise, among others) are not considered "deforestation."
b. Units of forest land subject to this "regeneration" process are successively allocated to forest classes with a higher average carbon-stock density. As in the degradation case, the difference in average carbon-stock density between two contiguous classes should be at least 10 percent.
c. The black line describes variations in carbon stocks over time in above-ground and below-ground biomass.

Contribution to Livelihoods

Home to more than 30 million people, Congo Basin forests support the livelihoods of more than 75 million people, from more than 150 ethnic groups, who rely on the local natural resources for food, medicine, and other needs (CBD 2009). Modern humans have occupied and used the Basin for at least 50,000 years. Evidence of the pygmy culture, which has adapted particularly well to the forest, dates back 20,000 to 25,000 years. Even today, a large portion of the the population living in the Congo Basin forest is indigenous. In addition to those inhabitants, many others directly or indirectly rely on the forest for fuel, food, medicines, and other nontimber products.

For the Congo Basin's population, the forest is a major source of food. The contribution of forests to food security is very often overlooked, but rural communities in the Basin get a significant portion of protein and fats in their diets from the wildlife in forests and along forest edges (Nasi, Taber, and Van Vliet 2011). Similarly, many communities depend on forested watersheds and mangrove ecosystems to support the freshwater and coastal fisheries. In addition, many fruits, nuts, grubs, mushrooms, honey, and other edibles are produced by forests and trees. A 2011 income survey conducted by the Center for International Forestry Research (CIFOR) of some 6,000 households in the Basin confirms that, on average, families living in and around forests derive between one-fifth and one-fourth of their income from forest-based sources: forests provide a source of cash income with which to purchase food (Wollenberg et al. 2011). Traditional hunter–gatherers also have complex, multigenerational relationships with farmers, exchanging forest products for starch-rich foods and access to manufactured goods (CBFP 2005).

Nontimber forest products (NTFPs) provide food, energy, and cultural items. Examples of NTFPs include honey, wax, propolis, bush mango, pygeum, raffia, gum arabic, kola nuts, bamboo, and wild plums. The use of these and other NTFPs varies widely based on culture, socioeconomic status, forest access, markets, and price, and to an extent (particularly for bush meat), on the legality of their harvest. Their sale in local and export markets contributes significantly to the livelihoods of forest dwellers (Ruiz Pérez et al. 2000; Shackleton, Shanley, and Ndoye 2007; Ingram et al. 2012).

Logging Sector: A Major Contributor to National Economies

Industrial logging has become one of the most extensive use of land in Central Africa, with almost 450,000 square kilometers of forest currently under concession (about a quarter of the total lowland tropical forests); whereas, in comparison, 12 percent of the land area is protected. It is expected that industrial logging concessions will expand further. The portion of forest area designated for logging is particularly high in the Republic of Congo (74 percent) and the Central African Republic (44 percent) (figure 1.1).

The formal logging sector in Central Africa produces an average of 8 million cubic meters of timber every year, mostly used for exports to Europe and Asia. Gabon is the largest producer, followed by Cameroon and the Republic of Congo (de Wasseige et al. 2009). Despite the Democratic Republic of Congo's vast forest resources, which represent more than 60 percent of the total forest area in the Congo Basin, the country is the smallest producer in the Basin, with just 310,000 cubic meters of formal timber production (table 1.3). This is essentially a consequence of a few investments in industrial logging operations due to the protracted conflict in the Democratic Republic of Congo over the past decade as well as other barriers, such as a lack of infrastructure needed to facilitate logging processes.

After a period of slow growth over the past 15 years, the timber production from Central Africa was contracted by about 2–3 million cubic meters in 2008, as a result of the global financial crisis that affected the market for tropical timber.[3] This contraction was particularly pronounced in countries with large export volumes, such as Cameroon and Gabon (figure 1.2). Production has recovered

Figure 1.1 Total Land Area, Total Dense Forest Area, and Area under Industrial Logging Concessions in the Congo Basin in 2010 (hectares)

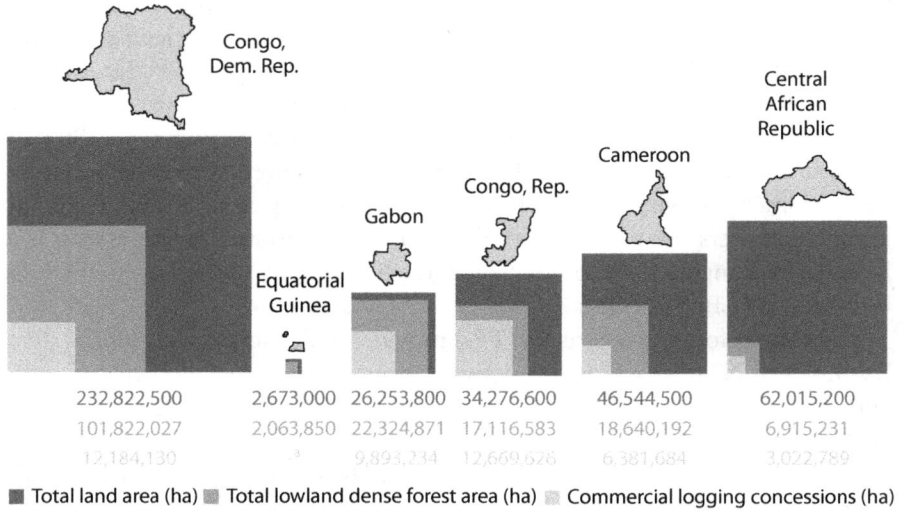

	Congo, Dem. Rep.	Equatorial Guinea	Gabon	Congo, Rep.	Cameroon	Central African Republic
Total land area (ha)	232,822,500	2,673,000	26,253,800	34,276,600	46,544,500	62,015,200
Total lowland dense forest area (ha)	101,822,027	2,063,850	22,324,871	17,116,583	18,640,192	6,915,231
Commercial logging concessions (ha)	12,184,130	a	9,893,234	12,669,626	6,381,684	3,022,789

■ Total land area (ha) ▒ Total lowland dense forest area (ha) ▒ Commercial logging concessions (ha)

Source: Authors, prepared from data in de Wasseige et al. 2012.
Note: ha = hectare.
a. In Equatorial Guinea, all logging concessions were cancelled in 2008.

Table 1.3 Harvested Timber Volume and Primary Species Logged by Country in 2006

Country	Production (m³)	Main species logged
Cameroon	2,296,254	*Ayous, sapelli, tali, azobé, iroko*
Central African Republic	537,998	*Ayous, sapelli, aniegré, iroko, sipo*
Congo, Dem. Rep.	310,000	*Sapelli, wengué, sipo, afromosia, iroko*
Congo, Rep.	1,330,980	*Sapelli, sipo, bossé, iroko, wengué*
Equatorial Guinea	524,799	*Okoumé, tali, azobé, ilomba*
Gabon	3,350,670	*Azobé, okan, movingui, ozigo*
Total	8,350,701	

Source: de Wasseige et al. 2009.
Note: m³ = cubic meter.

since then—in part, due to the steep increase in round wood production in Gabon toward the end of 2009.

The industrial logging sector remains one of the major contributors to the gross domestic product (GDP) of most Congo Basin countries. Historically, the forest sector has played an even more important role in the Congo Basin; however, with the booming development of the oil sector in several Basin countries over the last decade, the forest sector's relative contribution to overall GDP[4] has decreased. There is evidence that projected declines in oil production in Gabon over the next decade may lead to renewed growth in logging for export. Tax revenue from the forest sector in absolute terms is currently highest in Cameroon and Gabon, both of which are countries with well-developed commercial forestry sectors (table 1.4).

Figure 1.2 Annual Round Wood Production (m³) in the Congo Basin Countries

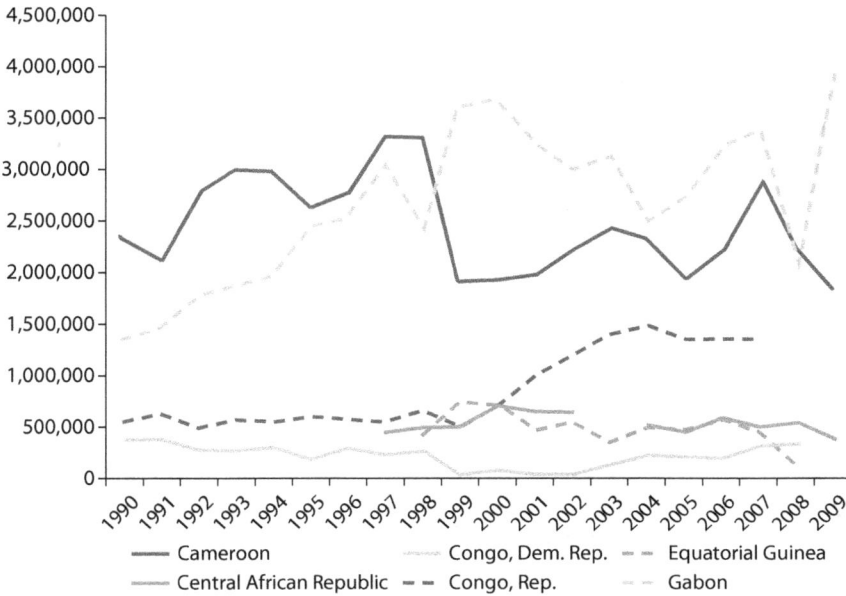

Source: de Wasseige et al. 2012.
Note: m³ = cubic meter.

Table 1.4 Contribution of the Forestry Sector to GDP and Gross Value Added, 2009

			Gross value added (US$ million)			
Country	Contribution to GDP		Roundwood production	Wood processing	Pulp and paper	Total for the forestry sector
	%	Year				
Cameroon	6	2004	236	74	13	324
Central African Republic	6.3	2009	133	10	1	144
Congo, Dem. Rep.	1	2003	185	2	—	186
Congo, Rep.	5.6	2006	45	27	—	72
Equatorial Guinea	0.2	2007	86	2	—	87
Gabon	4.3	2009	171	118	—	290
Total			856	233	14	1,103

Source: Authors, prepared with data on gross domestic product (GDP) from de Wasseige et al. 2009, and data on gross value added from FAO 2011.
Note: — = not available.

The industrial logging sector is also a vital employer, particularly in rural forested areas (FAO 2011). The formal timber sector accounts for about 50,000 full-time jobs in the six countries (table 1.5). Employment created in the formal forestry sector by private sector operators is particularly important in Gabon, where timber is the largest employment sector after the government. In Gabon,

Table 1.5 Direct Employment in Commercial Forest Production and Processing, 2006

| Country | Employment (1,000 FTE) | | | | Total forestry sector (% to total labor force) |
	Roundwood production	Wood processing	Pulp and paper	Total for the forestry sector	
Cameroon	12	8	1	20	0.3
Central African Republic	2	2	—	4	0.2
Congo, Dem. Rep.	6	—	—	6	—
Congo, Rep.	4	3	—	7	0.5
Equatorial Guinea	1	—	—	1	0.5
Gabon	8	4	—	12	1.9
Total	33	17	1	50	

Source: Authors, prepared with data from FAO 2011.
Note: — = not available; FTE = full-time equivalent.

the sector further provides indirect employment for another 5,000 jobs, and the public forest service itself employs about 600 officers and support staff. In Cameroon, the formal sector is estimated to have supplied some 20,000 full-time jobs in 2006; recent statistics from the government of Cameroon indicate that indirect employment from the sector could exceed 150,000 jobs (MINFOF-MINEP 2012).

The informal timber sector has long been overlooked but is now recognized as a major component of the logging industry. The recovery and boom of the domestic market in recent years is a sharp turnaround from its contraction following the 1994 devaluation of the regional currency (African Financial Community Franc, or FCFA) that boosted formal timber exportation at the expense of the local markets. Demand for timber has been soaring on local markets to meet the growing needs of urban populations for construction lumber and woodfuel/charcoal. It was also recently documented that well-established transnational timber supply networks from Central Africa to as far as Niger, Chad, Sudan, the Arab Republic of Egypt, Libya, and Algeria have developed, driving the growing urban demand for construction material (Langbour, Roda, and Koff 2010).

The domestic and regional timber economy is just as important as the formalized sector; in fact, in some countries the potential economic importance of the domestic forest economy appears to exceed the formal economy. In Cameroon, for example, domestic timber production already surpasses formal timber production, and in the Democratic Republic of Congo and the Republic of Congo, domestic timber production is estimated to represent more than 30 percent of the total timber production. Only recently, research on the informal sector substantiated the magnitude of the informal sector both in terms of estimated timber volumes and in the number of jobs related to informal activities (from production to marketing; Cerutti and Lescuyer 2011; Lescuyer et al. 2011). The informal sector is particularly significant from a local development perspective, as it provides for much higher direct and indirect local employment than does the formal sector, with benefits more equally redistributed at the local level than those achieved through formal sector activities (Lescuyer et al. 2012).

Progress on Sustainable Forest Management

Over the past two decades, Congo Basin countries have engaged in policies of sustainable forest management and conservation. After the Earth Summit in Rio de Janeiro in 1992, all of the Basin countries revised their forest laws in order to bring them into compliance with sustainable forest-management practices. The Forestry Commission of Central Africa (COMIFAC[5]) was founded to provide political and technical guidance, coordination, harmonization, and decision making in conservation and sustainable management of forest ecosystems and savannas in the region. In February 2005, during the second Summit of Heads of State in Brazzaville, COMIFAC adopted a plan, the "Plan de Convergence," for better forest management and conservation in Central Africa (box 1.2). At the September 2002 World Summit on Sustainable Development in Johannesburg, South Africa, the countries of the Congo Basin joined with partners from developed countries to create the Congo Basin Forest Partnership (CBFP). The CBFP funding is used to establish new national parks, strengthen governmental forest authorities, and provide opportunities for sustainable development.

The political commitment of Congo Basin countries to sustainable forest management, along with the support from the international community, has translated into significant progress in the following areas:

- **Protected areas**[6]: Major advancements have been made in terms of establishing protected areas. The primary function of these forests often is

---

Box 1.2 COMIFAC's "Plan de Convergence"

In 2005, COMIFAC adopted a plan that defines common intervention strategies of states and development partners in the conservation and sustainable management of forest ecosystems and savannas in Central Africa. It is structured around 10 strategic activities:

1. Harmonization of forest and fiscal policies
2. Knowledge of the resource
3. Development of forest ecosystems and reforestation
4. Conservation of biological diversity
5. Sustainable development of forest resources
6. Development of alternative activities and poverty reduction
7. Capacity building, stakeholder participation, information, and training
8. Research and development
9. Development of funding mechanisms
10. Cooperation and partnerships

Source: http://www.comifac.org/plan-de-convergence.

---

the conservation of biological diversity, the safeguard of soil and water resources, or the conservation of cultural heritage. Increased capacities in the management of protected areas and forest conservation areas have helped to reduce pressure on biodiversity. As of 2011, 341 protected areas[7] were established in the six Basin countries: they cover nearly 60 million hectares, representing about 14 percent of the territory covered by the six countries. The highest number of protected areas and the largest proportion of national territory covered can be found in Cameroon and the Central African Republic. Apart from the protected areas in category VI (recreational hunting zones and hunting reserves), biodiversity management in Central Africa is dominated by 46 national parks covering about 18.8 million hectares. National parks constitute the bulk of protected areas in countries like Gabon, which has 13 national parks out of 17 protected areas, covering an area of 2.2 million hectares out of a total of 2.4 million hectares (de Wasseige et al. 2009).

Capacities to manage protected areas, while still insufficient, have improved over the past few years; partnerships with international and local nongovernmental organizations (NGOs) have been established in most countries and yield fruit in terms of biodiversity preservation.

• **Sustainable forest management (SFM) in logging concessions:** While there was an overall increase in the adoption of management plans in all three main tropical forest regions (that is, Latin America, Asia and Pacific, and Africa), the relative increase was particularly significant in Africa, predominantly in the Congo Basin. The trend for the development of management plans has been momentous, from 0 hectares managed in 2000, to more than 7.1 million hectares of forest concessions in the subregion managed in accordance with state-approved plans in 2005, to 11.3 million hectares in 2008, and to 25.6 million hectares in 2010 (see table 1.6). The most considerable

Table 1.6　Forest Management in the Congo Basin Countries, 2005–2010
(thousands of hectares)

Country	Total		Available for harvesting		With management plans		Certified		Sustainably managed	
	2005	2010	2005	2010	2005	2010	2005	2010	2005	2010
Cameroon	8,840	7,600	4,950	6,100	1,760	5,000	—	705	500	1,255
Central African Republic	3,500	5,200	2,920	3,100	650	2,320	—	—	186	—
Congo, Dem. Rep.	20,500	22,500	15,500	9,100	1,080	6,590	—	—	284	—
Congo, Rep.	18,400	15,200	8,440	11,980	1,300	8,270	—	1,908	1,300	2,494
Equatorial Guinea	—	—	—	—	—	—	—	—	—	—
Gabon	10,600	10,600	6,923	10,300	2,310	3,450	1,480	1,870	1,480	2,420
Total	61,840	61,100	38,733	40,580	7,100	25,630	1,480	4,483	3,750	6,169

Source: Based on data collected from Blaser et al. 2011.
Note: — = not available.

progress in the development of management plans has occurred in Cameroon with 5.34 million hectares of natural forest now covered by management plans (as of 2011), compared with 1.76 million hectares in 2005. Management plans are now also in place for about 3.45 million hectares of natural forest in Gabon. The number of logging concessions with approved management plans is expected to increase further in the next five years, as a large part of the remaining concessions are currently preparing their management plans. Similarly, the area of certified natural forest production in the permanent forest estate in Central Africa has increased from just 1.5 million hectares certified in 2005 in Gabon to 4.5 million hectares certified in 2010 in Gabon, Cameroon, and the Republic of Congo (Blaser et al. 2011).

- **Illegal logging and forest governance**: Illegal logging is suspected to be widespread in the region, but little data exist to adequately quantify the scope. Annual losses in revenues and assets due to illegal logging on public lands are estimated at about US$10–18 billion worldwide, with losses mainly present in developing countries. In Cameroon, yearly losses are estimated at US$5.3 million; in the Republic of Congo, US$4.2 million; and in Gabon, $10.1 million. This revenue is lost every year from poor regulation of timber production, and the figures do not include estimates for "informal" logging carried out by small-scale operators, which mainly operate illegally. However, reliable figures on the volume of illegal logging are rarely available and differ greatly. The actual forest area affected is even more difficult to detect and delineate with current remote-sensing techniques, as unlawful logging in the Basin is usually associated with forest degradation rather than deforestation.

  Most Congo Basin countries have adhered to the European Union's FLEGT process. FLEGT stands for "Forest Law Enforcement, Governance, and Trade" and has been set up to strengthen forest governance and combat illegal logging. Cameroon (2010), the Republic of Congo (2010), and the Central African Republic (2011) have signed Voluntary Partnership Agreements (VPAs) negotiated under the European Union's FLEGT process. The FLEGT seeks to ban illegal timber trading on the European market. One of the fundamental elements of FLEGT is to provide support to timber-producing countries so they can improve their forest governance and establish effective methods to counter illegal logging (box 1.3). As of April 2012, six countries were developing the systems agreed to under a VPA, including Cameroon, the Central African Republic, the Republic of Congo, and four countries were negotiating with the EU, including the Democratic Republic of Congo and Gabon.

Deforestation and Forest Degradation

Overall Low Rates of Deforestation and Forest Degradation
The forest cover changes in the Congo Basin are among the lowest in the world's tropical rainforest belt; net deforestation rates are more than two times higher

Box 1.3 European Union's Forest Law Enforcement, Governance, and Trade Program

The European Union's Forest Law Enforcement, Governance, and Trade (FLEGT) action plan attempts to harness the power of timber-consuming countries to reduce the extent of illegal logging. The role of consumer countries in driving the demand for timber and wood products—and thereby contributing to illegal logging—has been a particular focus of debate in recent years. The problem has been especially relevant to the European Union (EU), which is a major global importer of timber and wood products; several countries from which EU member states import such products suffer from extensive illegal activities. Spurred by discussions at the East Asia FLEGT conference in September 2001, the European Commission published FLEGT in May 2003. Approved by the Council of the EU in October 2003, it included the following proposals:

- Support to timber-exporting countries, including action to promote equitable solutions to the illegal logging problem.
- Activities to promote trade in legal timber, including plans to develop and implement VPAs between the EU and timber-exporting countries.
- Promotion of public procurement policies, including action to guide contracting authorities on how to deal with legality when specifying timber during procurement.
- Support for private-sector initiatives, including action to encourage such initiatives for good practice in the forest sector, including the use of voluntary codes of conduct for private companies to source legal timber.
- Safeguards for financing and investment, including action to encourage banks and financial institutions investing in the forest sector to develop due-care procedures when granting credits.
- Use of existing legislative instruments or adaption of new legislation to support the plan (for example, the EU Illegal Timber Regulation).
- Addressing the problem of conflict timber.

For more information, see http://www.euflegt.efi.int/portal/home/flegt_intro/flegt_action_plan.

---

in South America and four times higher in Southeast Asia. In comparison, Brazil is estimated to have lost 0.5 percent of its forests per year (that is, about 28,000 square kilometers per year) over the past 20 years and Indonesia 1.0 percent per year (12,000 square kilometers per year) for the same period (FAO 2011). In other words, Brazil and Indonesia currently lose more forest in 2 years and 4 years, respectively, than all Basin countries did over the past 15 years. Figure 1.3 below indicates contribution by major blocks to the global loss of humid tropical forest cover during the 2000–05 period. Africa has contributed only 5.4 percent of the estimated global loss of humid tropical forest cover over the 2000–2005 period, compared with 12.8 percent in Indonesia and 47.8 percent in Brazil alone.

Deforestation rates in the Central African countries[8] are the lowest in Sub-Saharan Africa. Table 1.7[9] presents results from an analysis at a global level.

Figure 1.3 Contribution of Region to Humid Forest Loss across Regions

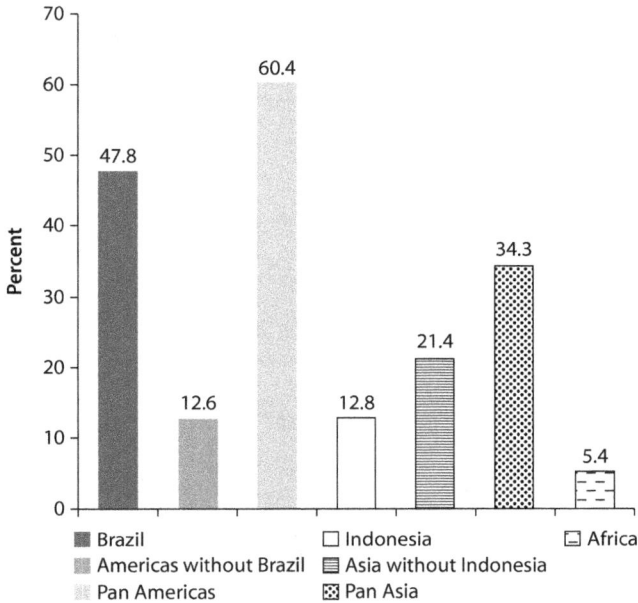

Source: Authors, derived from Hansen et al. 2008.

Overall figures confirm that the deforestation rates in Central Africa are not only well below those of the major negative contributors to world total forest area but also below the deforestation rates experienced by most other African regions. With regard to area, Central Africa loses about 40 percent less forest each year than southern Africa, 25 percent less than West Africa, and 15 percent less than east Africa, and represents less than one-fifth of the total forest area lost every year on the continent.

Deforestation trends are even lower for the highly forested Congo Basin countries. The overall annual net deforestation rate in the Basin rainforest was estimated at 0.09 percent during the 1990–2000 period. Over the 2000–05 period, this rate almost doubled, corresponding to a net loss of about 300,000 square kilometers each year. As shown in figure 1.4 and table 1.8, while deforestation rates have stabilized in the Central African Republic and even dropped in Gabon and Equatorial Guinea and Cameroon, they have significantly increased in the Republic of Congo and the Democratic Republic of Congo (Ernst et al. 2010).

Forest degradation—though harder to quantify—also drives major change in Congo Basin forests. As for deforestation, global forest degradation in the Basin has amplified in recent years. This trend is essentially driven by the Democratic Republic of Congo, as degradation rates have globally stabilized in the other countries. Gabon even shows a negative rate of degradation that indicates a global increase in forest density (table 1.9). In fact, the rate of net degradation in

Table 1.7 Changes in Forest Area in Africa and in the Main Negative Contributors[a] to World Total Forest Area, 1990–2010

Subregion	Forest area (thousand ha)			Annual change (thousand ha)		Annual change rate (%)	
	1990	2000	2010	1990–2000	2000–10	1990–2000	2000–10
Central Africa	268,214	261,455	254,854	−676	−660	−0.25	−0.26
East Africa	88,865	81,027	73,197	−784	−783	−0.92	−1.01
North Africa	85,123	79,224	78,814	−590	−41	−0.72	−0.05
Southern Africa	215,447	204,879	194,320	−1,057	−1,056	−0.50	−0.53
West Africa	91,589	81,979	73,234	−961	−875	−1.10	−1.12
Total Africa	749,238	708,564	674,419	−4,067	−3,414	−0.56	−0.49
Southeast Asia	247,260	223,045	214,064	−2,422	−898	−1.03	−0.41
Oceania	198,744	198,381	191,384	−36	−700	−0.02	−0.36
Central America	96,008	88,731	84,301	−728	−443	−0.79	−0.51
South America	946,454	904,322	864,351	−4,213	−3,997	−0.45	−0.45
World	4,168,399	4,085,063	4,032,905	−8,334	−5,216	−0.20	−0.13

Source: FAO 2011.
Note: For the purpose of this analysis:
Central Africa: Burundi, Cameroon, Central African Republic, Chad, Democratic Republic of Congo, Equatorial Guinea, Gabon, Republic of Congo, Rwanda, St. Helena, Ascension and Tristan da Cunha, São Tomé and Príncipe
East Africa: Comoros, Djibouti, Eritrea, Ethiopia, Kenya, Madagascar, Mauritius, Mayotte, Réunion, Seychelles, Somalia, Tanzania, Uganda
North Africa: Algeria, Arab Republic of Egypt, Libya, Mauritania, Morocco, Sudan, Tunisia, Western Sahara
Southern Africa: Angola, Botswana, Lesotho, Malawi, Mozambique, Namibia, South Africa, Swaziland, Zambia, Zimbabwe
West Africa: Benin, Burkina Faso, Cape Verde, Côte d'Ivoire, The Gambia, Ghana, Guinea, Guinea-Bissau, Liberia, Mali, Niger, Nigeria, Senegal, Sierra Leone, Togo
Southeast Asia: Brunei Darussalam, Cambodia, Indonesia, Lao People's Democratic Republic, Malaysia, Myanmar, Philippines, Singapore, Thailand, Timor-Leste, Vietnam
Oceania: American Samoa, Australia, Cook Islands, Federated States of Micronesia, Fiji, French Polynesia, Guam, Kiribati, Marshall Islands, Nauru, New Caledonia, New Zealand, Niue, Norfolk Island, Northern Mariana Islands, Palau, Papua New Guinea, Pitcairn, Samoa, Solomon Islands, Tokelau, Tonga, Tuvalu, Vanuatu, Wallis and Futuna Islands
Central America: Belize, Costa Rica, El Salvador, Guatemala, Honduras, Mexico, Nicaragua, Panama
South America: Argentina, Bolivia, Brazil, Chile, Colombia, Ecuador, Falkland Islands (Malvinas), French Guiana, Guyana, Paraguay, Peru, Suriname, Uruguay, República Bolivariana de Venezuela
a. Main positive contributors include East Asia (especially China), Europe, North America (especially the United States of America) and South Asia (especially India).

dense forests is a combination of gross degradation and forest recovery. It is, however, worth noting that the quantified measure of degradation is based solely on significant detected change in forest cover and not on qualitative terms (that is, change in species composition).

Deforestation in the Congo Basin is linked to population density and associated subsistence activities expansion (agriculture and energy) that usually happen at the expense of the forest. This is a completely different picture than in Indonesia, Brazil, and other countries, where large-scale agricultural operations (for example, conversion to pasture and plantations) are by far the main drivers of deforestation[10] (Kissinger 2011). Urban centers in the Basin[11] are growing rapidly at 3–5 percent per year and even more (5–8 percent) for the already large cities, such as Kinshasa and Kisangani; Brazzaville and Pointe Noire; Libreville,

Figure 1.4 Changes in Forest Area in Main Regions in Africa on 1990–2010 period (in million hectares)

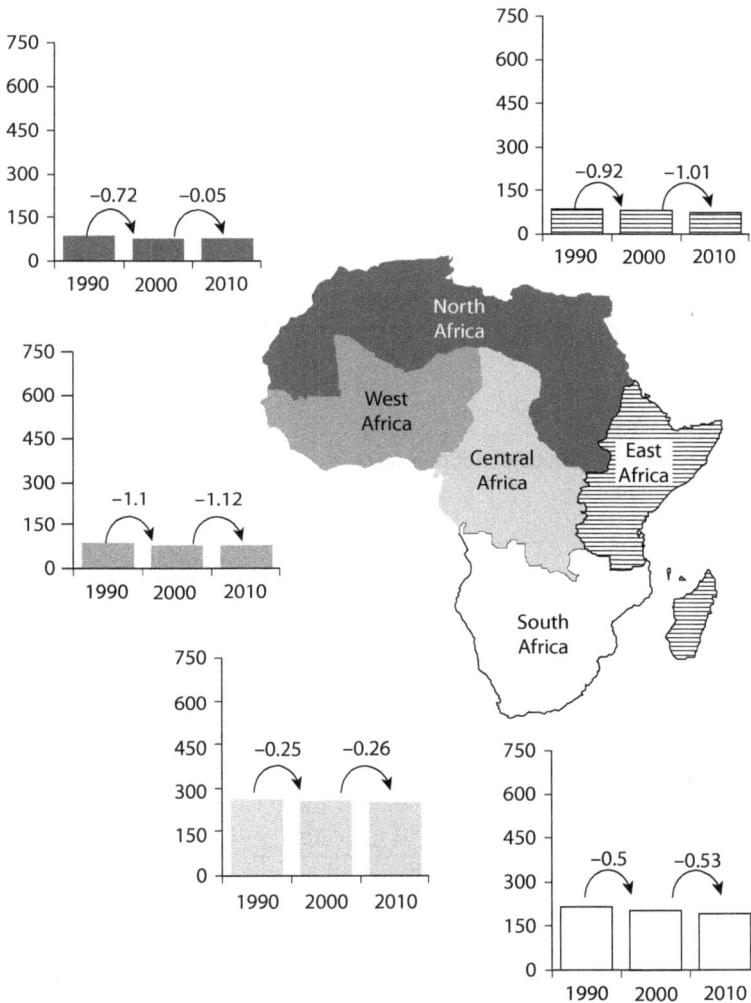

Source: Authors, based on FAO 2011.

Franceville, and Port Gentil; Douala and Yaounde; and Bata. Consequently, deforestation and forest degradation are currently mainly concentrated around urban centers and in the most densely populated areas.

Logging activities lead to forest degradation rather than deforestation. Unlike in other tropical regions, where logging activities usually entail a transition to another land use, logging in the Basin is highly selective and extensive, and production forests remain permanently forested. In industrial concession, wood extraction is very low, with an average rate of less than 0.5 cubic meters per

Table 1.8 Congo Basin Countries' Net Annual Deforestation Rates in Dense Forests during 1990–2000 and 2000–05

Percent

	Net annual deforestation	
	1990–2000	*2000–05*
Cameroon	0.08	0.03
Central African Republic	0.06	0.06
Congo, Dem. Rep.	0.11	0.22
Congo, Rep.	0.03	0.07
Equatorial Guinea	0.02	—
Gabon	0.05	0
Total Congo Basin	0.09	0.17

Source: de Wasseige et al. 2012.
Note: — = not available.

Table 1.9 Congo Basin Countries' Net Annual Degradation Rates in Dense Forests during 1990–2000 and 2000–05

Percent

	Net annual degradation	
	1990–2000	*2000–05*
Cameroon	0.06	0.07
Central African Republic	0.03	0.03
Congo, Dem. Rep.	0.06	0.12
Congo, Rep.	0.03	0.03
Equatorial Guinea	0.03	—
Gabon	0.04	−0.01
Total Congo Basin	0.05	0.09

Source: de Wasseige et al. 2012.
Note: — = not available.

hectare. This outcome results from the highly selective method applied in the Congo Basin (box 1.4).

A Mostly "Passive" Protection

Deforestation and degradation in the Congo Basin have been limited through a combination of various factors: low population densities, historical political instability, poor infrastructure, and a nonconducive business environment for private sector investment. These factors have created a kind of "passive protection" of the forest. Moreover, oil or other natural resources booms and Dutch disease effects[12] in some of the Basin countries spurred wages and created jobs in urban areas, stimulating rural–urban migration.

- **Average rural population density is very low**: Although the total population of the six countries is estimated to be about 96 million people in 2010 (World

---

Box 1.4 Highly Selective Logging Activities in the Congo Basin

Of the more than 100 species generally available in the tropical humid forest in Central Africa, fewer than 13 are usually harvested. Further, the three most harvested species (*okoumé, sapelli,* and *ayous*) combined represent about 59 percent of log production in Central Africa. Though the countries would like to see more secondary species logged in the Congo Basin forests, the export markets have thus far shown themselves to be conservative and slow in accepting unfamiliar secondary species, regardless of their otherwise perfectly suitable technological characteristics. In general, selectivity in logging increases when the harvesting costs are high, because timber companies tend to concentrate only on the most economically rewarding species. Nevertheless, the number of species logged is gradually diversifying, yet, thus far, only in forests near the ports of export and other areas with lower production costs (for example, Cameroon, coastal areas of Gabon, southern Congo, and the province of Bas-Congo in the Democratic Republic of Congo; de Wasseige et al. 2012).

Bank 2012), the Congo Basin itself is sparsely populated, with an estimated 30 million people, more than half in urban areas (including 9 million people in Kinshasa). Average rural population density is therefore very low, estimated at 6.5 inhabitants per square kilometer, with densities as low as one to three people per square kilometer in the Congo River central cuvette. Some zones in central and northeast Gabon, northern Republic of Congo, and the Democratic Republic of Congo are reported to be part of the 10 percent wildest zones on earth.[13] Despite high demographic rates, population densities in forested areas have remained low because of steady rural–urban migration.

- **Political instability in the region over the last 20 years has paralyzed economic developments**. The Central African Republic, for example, has experienced numerous rebellions and sporadic conflicts, resulting in the exodus of nearly 300,000 people. A similar situation is seen in the Republic of Congo. The country suffered an armed conflict between 1997 and 2003, which impoverished the country and caused considerable damage to the infrastructure and the national economy. In the Democratic Republic of Congo, state mismanagement of resources for 30 years, sporadic looting, and two periods of armed conflict have destroyed much of the infrastructure and have led to institutional collapse.

- **Poor infrastructure in the Congo Basin has hampered the development of national and regional economies**. It has been particularly restricting for the agriculture sector, constraining the transition from subsistence agriculture to a more market-oriented one. Inadequate market access throughout the region has made any transition to a more intensive form of agriculture next to impossible. Feeder roads in the humid forest are difficult to maintain under wet conditions; in many cases, they are impassable during the rainy season. In the

Democratic Republic of Congo, river transport proves to be one of the most efficient means of conveyance; however, it only works intermittently, depending on the water level. Furthermore, limited storage and processing capacities prevent farmers from waiting for the dry season to access markets and sell their products. Most farmers are therefore completely isolated from potential venues to sell their products and purchase inputs and thereby are cut off from participation in the broader economy that could foster competition and growth. Poor road infrastructure combined with administrative difficulties (proliferation of roadblocks, in particular) has been a major obstacle in developing regional trade.

The same situation applies to the mining sector, where adequate infrastructure was considered a prerequisite to any new investments in mining operations; however, with the high demand for minerals and the high prices, incentives to develop new mineral deposits increase with new deals. In fact, over the past few years, a trend toward investors offering to build associated infrastructures—roads, railways, power plants, large dams, and ports—has emerged. These new deals largely remove the burden from host countries that generally lack the financial capacities to cover the investment. This trend would circumvent one of the major weaknesses for developing mining operations of the Congo Basin countries.

- **Finally, poor governance and a lack of clear regulatory frameworks have discouraged private investments.** Complex and often arbitrary and predatory taxation rules in the Basin countries (World Bank 2010), combined with a climate of instability and a lack of governance and clear laws, resulted in investment capital flows out of the region. In addition, the heavy reliance of some of these economies on oil took the governments' focus away from the need for economic diversification.

The Congo Basin region has not witnessed the expansion of large-scale plantation that other tropical regions have seen. The Basin does have significant agro-ecological potential for the growth of several major commodities, including soybeans, sugarcane, and oil palm. However, the weak transport network, low land productivity records, and poor business environment overall are major weaknesses that deter investors. Given the availability of suitable land for agricultural expansion in other countries as well as better performance in terms of infrastructure and productivity, and an enabling business environment, the Congo Basin has not attracted sizeable investment in large-scale agriculture until now.

An "HFLD Profile" for Congo Basin Countries

"Forest transition" is used to describe a sequence in forest cover. The transition curve, a concept introduced by Mather (1992), indicates patterns that could apply to a forested country as it progresses on its developmental curve. There is some evidence that the forest cover of a country shrinks as the country develops and pressures on natural resources increase. According to the forest transition

(FT) theory (box 1.5), countries early in their economic development are characterized by high forest cover and low deforestation (HFLD). Then, deforestation tends to increase with time and economic development until a certain minimum cover of forest has been reached. Eventually, according to the FT theory, countries slow deforestation and forest cover boosts again, normally in parallel with a more diversified economy that is less dependent on forest, land, and other natural resources for wealth and employment generation. While the FT theory does not presume specific predictions, for many countries it indicates the correlation between development and forest cover of a country or region.

Most Congo Basin countries are still located in the first stage (little disturbance) of the FT, but there are signs that the Congo Basin forest is under increasing pressure from a variety of forces, including oil and mineral extraction,

---

Box 1.5 Forest Transition Theory: Where Do Congo Basin Countries Stand?

Congo Basin countries are still located in the first stage of the forest transition (FT), with a high forest cover and low deforestation (HFLD) profile. Countries at the second stage, such as parts of Brazil, Indonesia, and Ghana, have large tracts of forest with high deforestation rates at the forest frontier, mainly as a result of expansion of cropland and pasture in combination with colonization. Countries at stage three have low deforestation and low forest cover characterized by forest mosaics and stable forest areas. In these countries, such as India, deforestation rates, though initially high, have leveled off because forests have been largely cleared and protection policies have been put in place. Other countries have reached stage four, such as China and Vietnam, where they are increasing their forest cover through afforestation and reforestation.

Figure B1.5.1 The Forest Transition Theory: Where Do the Congo Basin Countries Stand?

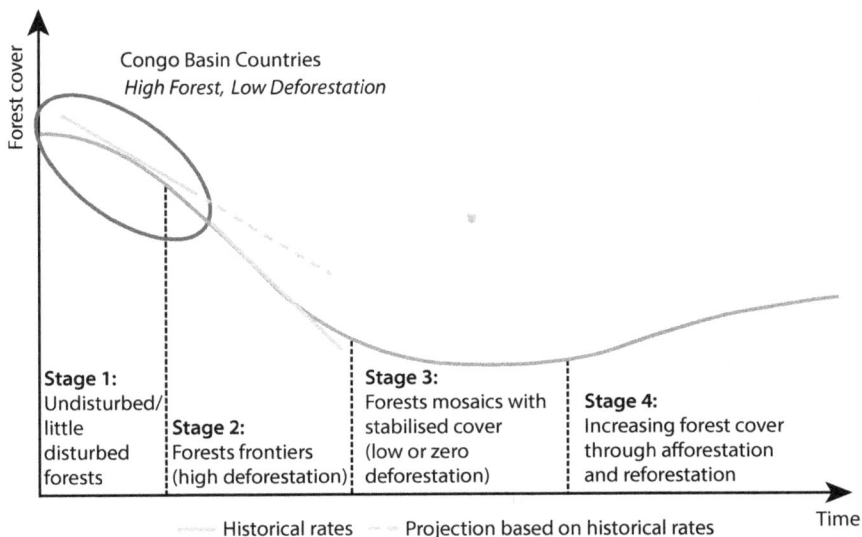

Congo Basin Countries
High Forest, Low Deforestation

Forest cover

Stage 1: Undisturbed/ little disturbed forests

Stage 2: Forests frontiers (high deforestation)

Stage 3: Forests mosaics with stabilised cover (low or zero deforestation)

Stage 4: Increasing forest cover through afforestation and reforestation

Historical rates   Projection based on historical rates

Time

---

road development, agribusiness, biofuels, agricultural expansion for subsistence, and population growth. Such factors might drastically amplify the rate of deforestation and forest degradation in the coming years and initiate the transition to second stage of the forest curve (forest frontiers).

The FT curve is by no means prescriptive, and leapfrogging is possible. In the context of low-carbon development strategies and associated financing instruments—specifically the REDD+ mechanism[14]—the Congo Basin countries are seeking to develop strategies that would allow them to leapfrog the "high deforestation" stage and head directly to a expansion scheme that would limit negative impacts on natural forests. Such a leapfrogging would require an in-depth understanding of the existing and future pressures on forests as well as an ambitious set of policy reforms that could gear a more "forest-friendly" development path.

Notes

1. Some scientists regard this classification system as unsatisfactory because not all of the boundaries between the ecoregions correspond to the reality on the ground (CBFP 2006); however, many conservation NGOs use the concept of ecoregions, among others, as a tool for planning further research.

2. Plants absorb carbon dioxide through photosynthesis and release some of it through respiration and decomposition; the remainder is stored in biomass, necromass, and soil.

3. OFAC. National Indicators. 2011. www.observatoire-comifac.net, Kinshasa (accessed March 2012).

4. The contribution of the forestry sector to GDP has decreased gradually and consistently, particularly for countries with a growing oil sector, notably the Republic of Congo, Gabon, and Equatorial Guinea. For Equatorial Guinea, in particular, the forest sector's contribution to GDP dropped from 17.9 percent in 1990 to 0.9 percent in 2006 (FAO 2011).

5. Established in March 1999, COMIFAC establishes a working platform for 10 countries in Central Africa (Burundi, Cameroon, the Central African Republic, Chad, the Republic of Congo, Gabon, the Democratic Republic of Congo, Equatorial Guinea, Rwanda, and São Tomé and Príncipe).

6. This section refers to all protected areas in the Basin countries and does not distinguish between forested and nonforested protected areas.

7. As per the International Union for Conservation of Nature (IUCN) Category I to VI. The exact distribution of protected areas between different IUCN categories is difficult given the differences in conceptions between the stakeholders and the laws of different countries.

8. Note that Central Africa refers to more countries than the six highly forested countries covered by this study.

9. The data presented in this table have been extracted from the FAO (2011). It should be highlighted that FAO data differ from Congo Basin–specific data put together by the Observatory for the Forests of Central Africa (OFAC) and presented in the reports *State of Forests in Congo Basin* (de Wasseige et al. 2008 and 2010 editions).

While the authors relied on FAO statistics for global data on forests, they used OFAC statistics for Congo Basin–specific data.

10. "Industrial soybean cultivation accounts for 70 percent of Argentina's deforestation while Vietnam's export commodities of coffee, cashew, pepper, shrimp (the latter affecting coastal mangroves), rice, and rubber drive forest conversion. Other countries, with significant commercial and industrial impacts on forests include: the Lao People's Democratic Republic (plantation fueled by foreign direct investment), Costa Rica (meat exports to the US promoted by government lending policies), Mexico (82 percent of deforestation due to agriculture or grazing), and Tanzania (increasing biofuel production)." (Kissinger 2011, 2).

11. Total population in the Congo Basin is projected to increase by about 70 percent through 2030.

12. This economic concept illustrates the relationship between the increase in exploitation of natural resources in a country and the respective decline of the manufacturing sector. An increase in revenues inflow from exports of natural resources will appreciate the country's currency, making its manufactured products more expensive for other countries. The manufacturing sector becomes less competitive than that of countries with weaker currencies. This phenomenon was common in postcolonial African states in the 1990s.

13. Using the "Human Footprint" approach as defined by Sanderson et al. (2002) (de Wasseige et al. 2009).

14. The REDD+ concept, as currently defined, embraces "reducing emissions from deforestation and forest degradation, and the role of conservation, sustainable management of forests, and enhancement of forest carbon stocks." The scope and design of the financing mechanism associated to REDD+ is still in negotiations under the UNFCCC auspices (see chapter 3).

References

Billand, A. 2012. "Biodiversity in Central African Forests: An Overview of Knowledge, Challenges, and Conservation Measures." In *The Forests of the Congo Basin—State of the Forest 2010*, ed. de Wasseige et al. Luxembourg: Publications Office of the European Union.

Blaser, J., A. Sarre, D. Poore, and S. Johnson. 2011. "Status of Tropical Forest Management 2011." ITTO Technical Series 38, International Tropical Timber Organization, Yokohama, Japan.

Canadel, J., and M. Raupach. 2008. "Managing Forests for Climate Change Mitigation." *Science* 320 (5882): 1456–57.

CBD (Convention on Biological Diversity). 2009. *Biodiversity and Forest Management in the Congo Basin: Ten Good Forest Management and Development Practices That Consider Biodiversity, Poverty Reduction and Development*. Montreal, Canada: CBD.

CBFP (Congo Basin Forest Partnership). 2005. *The Forests of the Congo Basin: A Preliminary Assessment*. Congo Basin Forest Partneship, Yaoundé, Cameroon.

———. 2006. *The Forests of the Congo Basin, State of the Forest 2006*. Congo Basin Forest Partneship, Yaoundé, Cameroon. http://www.giz.de/Themen/de/SID-E1A6CC9F-7E770AE7/dokumente/en-state-of-forests-congo-basin-2006.pdf.

Cerutti, P. O., and G. Lescuyer. 2011. "The Domestic Market for Small-Scale Chainsaw Milling in Cameroon: Present Situation, Opportunities and Challenges." Occasional Paper 61, Center for International Forestry Research (CIFOR), Bogor, Indonesia.

Chapin, F. S., J. Randerson, A. D. McGuire, J. Foley, and C. Field. 2008. "Changing Feedbacks in the Climate-Biosphere System." *Frontiers in Ecology and the Environment* 6: 313–20.

DeFries, R., R. Houghton, and M. Hansen. 2002. "Carbon Emissions from Tropical Deforestation and Regrowth Based on Satellite Observations for the 1980s and 90s." *Proceedings of the National Academy of Sciences (PNAS)* 99 (22): 14256–61. http://www.pnas.org_cgi_doi_10.1073_pnas.182560099.

de Wasseige, C., P. de Marcken, F. Hiol-Hiol, P. Mayaux, B. Desclee, R. Nasi, A. Billand, P. Defourny, and R. Eba'a Atyi, eds. 2012. *The Forests of the Congo Basin—State of the Forest 2010.* Luxembourg: Publications Office of the European Union.

de Wasseige, C., D. Devers, P. de Marcken, R. Eba'a Atyi, R. Nasi, and P. Mayaux, eds. 2009. *The Forests of the Congo Basin—State of the Forest 2008.* Luxembourg: Publications Office of the European Union.

Ernst, C., A. Verhegghen, C. Bodart, P. Mayaux, C. de Wasseige, A. Bararwandika, G. Begoto, F. Esono Mba, M. Ibara, A. Kondjo Shoko, H. Koy Kondjo, J. S. Makak, J. D. Menomo Biang, C. Musampa, R. Ncogo Motogo, G. Neba Shu, B. Nkoumakali, C. B. Ouissika, and P. Defourny. 2010. "Congo Basin Forest Cover Change Estimate for 1990, 2000 and 2005 by Landsat Interpretation Using an Automated Object-Based Processing Chain." *The International Archives of the Photogrammetry, Remote Sensing and Spatial Information Sciences* XXXVIII-4/C7.

Ernst, C., A. Verheggen, P. Mayaux, M. Hansen, and P. Defourny. 2012. "Central African Forest Cover and Cover Change Mapping." In *The Forests of the Congo Basin—State of the Forest 2010,* ed. de Wasseige et al. Luxembourg: Publications Office of the European Union.

Ervin, J., N. Sekhran, A. Dinu, S. Gidda, M. Vergeichik, and J. Mee. 2010. *Protected Areas for the 21st Century: Lessons from UNDP/GEF's Portfolio.* New York: United Nations Development Programme and Convention on Biological Diversity.

FAO (Food and Agriculture Organization of the United Nations). 2010. *Statistical Yearbook 2010.* Rome: FAO: http://www.fao.org/docrep/015/am081m/am081m00.htm.

———. 2011. *State of the World's Forests 2011.* Rome: FAO.

Hansen, M., S. Stehman, P. Potapov, T. Loveland, J. Townshend, R. Defries, K. Pittman, B. Arunarwati, F. Stolle, M. Steininger, M. Carroll, and C. DiMiceli. 2008. "Humid Tropical Forest Clearing from 2000 to 2005 Quantified by Using Multitemporal and Multiresolution Remotely Sensed Data." *Proceedings of the National Academy of Sciences (PNAS)* 105 (27): 9439–44.

Harris, N., S. Brown, S. Hagen, S. Saatchi, S. Petrova, W. Salas, M. Hansen, P. Potapov, and A. Lotsch. 2012. "Baseline Map of Carbon Emissions from Deforestation in Tropical Regions." *Science* 336 (6088): 1573–76.

Houghton, J. T., Y. Ding, D. J. Griggs, M. Noguer, P. J. van der Linden, X. Dai, K. Maskell, and C. A. Johnson, eds. 2001. *Climate Change 2001: The Scientific Basis. Contribution of Working Group I to the Third Assessment Report of the Intergovernmental Panel on Climate Change.* Cambridge, UK: Cambridge University Press.

Ingram, V., O. Ndoye, D. Iponga, J. Tieguhong, and R. Nasi. 2012. "Non-Timber Forests Products: Contribution to National Economies and Strategies for Sustainable Development." In *The Forests of the Congo Basin—State of the Forest 2010*, ed. de Wasseige et al. Luxembourg: Publications Office of the European Union.

Kissinger, G. 2011. "Linking Forests and Food Production in the REDD+ Context." CCAFS Policy Brief 3, CGIAR Research Program on Climate Change, Agriculture and Food Security, Copenhagen, Denmark.

Langbour, P., J.-M. Roda, and Y. A. Koff. 2010. "Chainsaw Milling in Cameroon: The Northern Trail." *European Tropical Forest Research Network News* 52: 129–37.

Lescuyer, G., P. O. Cerutti, E. Essiane Mendoula, R. Eba'a Atyi, and R. Nasi. 2012. "An Appraisal of Chainsaw Milling in the Congo Basin." In *The Forests of the Congo Basin—State of the Forest 2010*, ed. de Wasseige et al. Luxembourg: Publications Office of the European Union.

Lescuyer, G., P. O. Cerutti, S. N. Manguiengha, and L. B. bi Ndong. 2011. "The Domestic Market for Small-Scale Chainsaw Milling in Gabon: Present Situation, Opportunities and Challenges." Occasional Paper 65, CIFOR, Bogor, Indonesia.

Mather, A. 1992. "The Forest Transition." *Area* 24 (4): 367–79.

MINFOF-MINEP (Cameroon, Ministry of the Forests and Ministry of Environment). 2012. "Employees in Forestry Sector in 2004." MINEF. http://data.cameroun-foret .com/livelihoods/employees-forestry-sector.

Nasi, R., P. Mayaux, D. Devers, N. Bayol, R. Eba'a Atyi, A. Mugnier, B. Cassagne, A. Billand, and D. Sonwa. 2009. "A First Look at Carbon Stocks and Their Variation in Congo Basin Forests." In *The Forests of the Congo Basin—State of the Forest 2008*, ed. de Wasseige et al. Luxembourg: Publications Office of the European Union.

Nasi, R., A. Taber, and N. Van Vliet. 2011. "Empty Forests, Empty Stomachs? Bushmeat and Livelihoods in the Congo and Amazon Basins." *International Forestry Review* 13 (3): 355–68.

Olson, D., and E. Dinerstein. 2002. "The Global 200: Priority Ecoregions for Global Conservation." *Annals of Missouri Botanical Garden* 89 (2): 199–224.

OFAC (Observatory for the Forests of Central Africa). 2011. "National Indicators." Accessed in March 2012. www.observatoire-comifac.net.

Ruiz Pérez, M., O. Ndoye, A. Eyebe, and A. Puntodewo. 2000. "Spatial Characteristics of Non-timber Forest Products Markets in the Humid Forest Zone of Cameroon." *International Forestry Review* 2 (2): 71–83.

Sanderson, E., M. Jaiteh, M. Levy, K. Redford, A. Wannebo, and G. Woolmer, 2002. "The Human Footprint and the Last of the Wild." *BioScience* 52 (10): 891–904.

Shackleton, S., P. Shanley, and O. Ndoye. 2007. "Invisible but Viable: Recognising Local Markets for Non-timber Forest Products." *International Forestry Review* 9 (3): 697–712.

West, P. C., G. Narisma, C. Barford, C. Kucharik, and J. Foley. 2011. "An Alternative Approach for Quantifying Climate Regulation by Ecosystems." *Frontiers in Ecology and the Environment* 9 (2): 126–33.

Wollenberg, E., B. M. Campbell, P. Holmgren, F. Seymour, L. Sibanda, and J. von Braun. 2011. "Actions Needed to Halt Deforestation and Promote Climate-Smart Agriculture." CCAFS Policy Brief 4, CGIAR Research Program on Climate Change Agriculture and Food Security, Copenhagen, Denmark.

World Bank. 2010. *Doing Business 2010*. Washington, DC: World Bank. http://www.doingbusiness.org/.

———. 2012. World Development Indicators, World dataBank on Health Nutrition and Population Statistics HNPS. http://databank.worldbank.org/ddp/home.do. World Bank, Washington, DC.

WWF (World Wildlife Fund). 2012. *World Wildlife Fund for Nature, Forests of the Green Heart of Africa*. Washington, DC: WWF. http://wwf.panda.org/what_we_do/where_we_work/congo_basin_forests/the_area/ecosystems_congo/forests.

CHAPTER 2

What Will Drive Deforestation in the Congo Basin? A Multisectoral Analysis

As chapter 1 indicates, despite historically low deforestation rates, Congo Basin countries are likely to enter a new phase of economic development that could lead to increased pressure on forests. In 2008, the Congo Basin countries, donors, and partner organizations agreed to pool their efforts to conduct a robust and scientifically sound study to thoroughly analyze deforestation and forest degradation in the Congo Basin. The main objective—to investigate in depth the major drivers of deforestation and forest degradation over the next decades in the Basin—specifically aimed at devising a regional economic model that will help building different scenarios of potential impacts of economic activities on forest cover in the next 20 to 30 years.

This chapter presents the chief findings of the research. It was conducted over the last two years in close consultation with the Basin countries and the regional Forestry Commission for Central Africa (COMIFAC). The study combined robust sectoral analysis of key economic sectors (transport, agriculture, logging, energy, and mining), a modeling exercise, and regular and iterative consultations with technical experts from the region. Chapter 2 is structured as follows: a first section will present the principle underlying the deforestation dynamics as well as the modeling approach adopted for the proposed exercise in the Congo Basin, and the second section will then summarize developments in the selected key economic sectors of the Congo Basin and their potential impacts on Basin forests.

Deforestation and Forest Degradation Dynamics in the Congo Basin

Proximate and Underlying Causes of Deforestation

Determining the driving forces of land use/cover change is complex. Human pressure on forests is shaped by a complex of market access, suitability for farming, and tenure security (Chomitz et al. 2007). According to the work from Geist and Lambin (2001), a combination of economic factors, institutions, national policies, and remote factors drive tropical deforestation (see figure 2.1).

Figure 2.1 Proximate and Underlying Causes of Deforestation and Forest Degradation

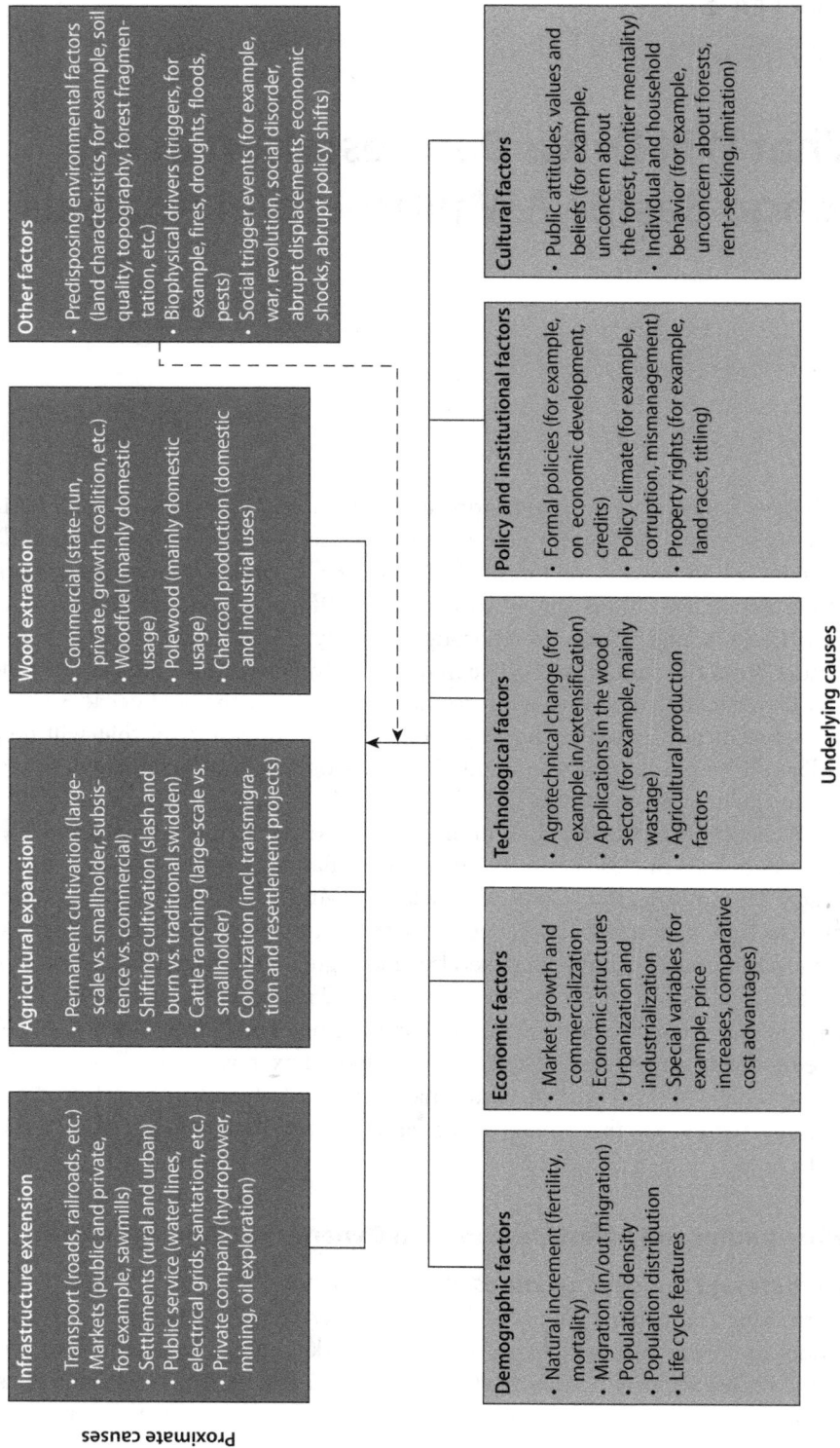

Proximate causes

Infrastructure extension
- Transport (roads, railroads, etc.)
- Markets (public and private, for example, sawmills)
- Settlements (rural and urban)
- Public service (water lines, electrical grids, sanitation, etc.)
- Private company (hydropower, mining, oil exploration)

Agricultural expansion
- Permanent cultivation (large-scale vs. smallholder, subsistence vs. commercial)
- Shifting cultivation (slash and burn vs. traditional swidden)
- Cattle ranching (large-scale vs. smallholder)
- Colonization (incl. transmigration and resettlement projects)

Wood extraction
- Commercial (state-run, private, growth coalition, etc.)
- Woodfuel (mainly domestic usage)
- Polewood (mainly domestic usage)
- Charcoal production (domestic and industrial uses)

Other factors
- Predisposing environmental factors (land characteristics, for example, soil quality, topography, forest fragmentation, etc.)
- Biophysical drivers (triggers, for example, fires, droughts, floods, pests)
- Social trigger events (for example, war, revolution, social disorder, abrupt displacements, economic shocks, abrupt policy shifts)

Underlying causes

Demographic factors
- Natural increment (fertility, mortality)
- Migration (in/out migration)
- Population density
- Population distribution
- Life cycle features

Economic factors
- Market growth and commercialization
- Economic structures
- Urbanization and industrialization
- Special variables (for example, price increases, comparative cost advantages)

Technological factors
- Agrotechnical change (for example in/extensification)
- Applications in the wood sector (for example, mainly wastage)
- Agricultural production factors

Policy and institutional factors
- Formal policies (for example, on economic development, credits)
- Policy climate (for example, corruption, mismanagement)
- Property rights (for example, land races, titling)

Cultural factors
- Public attitudes, values and beliefs (for example, unconcern about the forest, frontier mentality)
- Individual and household behavior (for example, unconcern about forests, rent-seeking, imitation)

Source: Geist and Lambin 2001.

- Proximate causes of deforestation are those human activities that typically operate at the local level and affect land use and impact forest cover. Commonly, these actions are grouped as agricultural expansion—such as shifting cultivation or cattle ranching—wood extraction (through logging or charcoal production), and infrastructure extension, including settlement expansion, transport infrastructure, or market infrastructure.

- Underpinning these proximate causes are underlying causes, a complex of economic issues, policies, and institutional matters; technological factors; cultural or sociopolitical concerns; and demographic factors. Other issues associated with deforestation are predisposing land characteristics (for example, slope and topography), features of the biophysical environment (soil compaction, drought conditions), and societal trigger events, such as social unrest or refugee movements.

Causes and drivers of tropical deforestation, however, cannot be reduced to a few variables. The interplay of several proximate as well as underlying factors drives deforestation in a synergetic way. The important underlying factors of policy and institutional factors—such as formal state policies, policy climate, and property rights arrangements—exert the strongest impact on proximate causes, while economic factors dominate the overall frequency pattern of cause occurrence (Geist and Lambin 2002).

Main Causes of Deforestation in the Congo Basin
In Central Africa, expansion of agricultural land is the most frequently reported proximate cause of tropical deforestation. Zhang et al. (2002) established with a geographic information system (GIS)-based assessment that subsistence small-scale farming was the principal determinant of deforestation in Central Africa, particularly along the edges between moist forests and nonforest land, where forests are more accessible. A detailed GIS-image interpretation confirmed the hypothesized relationship between deforestation and forest accessibility (Zhang et al. 2005): the extension of infrastructure, primarily road construction, also appears as an important proximate cause of deforestation and forest degradation in the Congo Basin (Duveiller et al. 2008). For example, the Douala-Bangui road from Cameroon to the Central African Republic, completed in 2003 and cutting across 1,400 kilometers in the northwestern Basin, has promoted massive logging, poaching, and forest loss (Laurance, Goosem, and Laurance 2009).

Demographic factors are a major underlying cause of deforestation in the Congo Basin, as they are chiefly associated with the growth of subsistence activities (agriculture and energy) and are thus strongly correlated with demographic patterns. As such, deforestation and forest degradation have been so far mainly concentrated around urban centers and in the most densely populated areas. Although Basin countries still have globally low population density rates, urbanization trends are emerging: urban centers in the Congo Basin are growing rapidly at 3–5 percent per year and even more so (5–8 percent) for the already large cities, such as Kinshasa and Kisangani, Brazzaville and Pointe Noire, Libreville, Franceville and Port Gentil, Douala and Yaounde, and Bata. Kinshasa alone is reported to have a population of 9 million. These growing urban centers create new dynamics and needs in terms of food and energy (mainly charcoal) supply, both of which are likely to be met by increased pressures on forest areas. Table 2.1 illustrates the population dynamics in the Basin countries; figure 2.2 shows the urbanization trend since 1995.

Rural areas in the rainforest are also becoming more densely populated, as evidenced by the proliferation of urban centers with a population of at least

Table 2.1 Rural/Urban Population and Urbanization Trends in the Congo Basin Countries

	1995	2000	2005	2010
Cameroon				
Total population	13,940,337	15,678,269	17,553,589	19,598,889
Population growth (%)	2.55	2.29	2.24	2.19
Urban population (% of total)	45.3	49.9	54.3	58.4
Urban population growth (%)	4.6	4.15	3.87	3.6
Central African Republic				
Total population	3,327,710	3,701,607	4,017,880	4,401,051
Population growth (%)	2.44	1.89	1.65	1.9
Urban population (% of total)	37.2	37.6	38.1	38.9
Urban population growth (%)	2.66	2.1	1.91	2.31
Congo, Dem. Rep				
Total population	44,067,369	49,626,200	57,420,522	65,965,795
Population growth (%)	3.27	2.44	2.94	2.71
Urban population (% of total)	28.4	29.8	32.1	35.2
Urban population growth (%)	3.69	3.38	4.39	4.48
Congo, Rep.				
Total population	2,732,706	3,135,773	3,533,177	4,042,899
Population growth (%)	2.74	2.6	2.51	2.54
Urban population (% of total)	56.4	58.3	60.2	62.1
Urban population growth (%)	3.48	3.25	3.14	3.16
Equatorial Guinea				
Total population	442,527	520,380	607,739	700,401
Population growth (%)	3.34	3.2	3	2.79
Urban population (% of total)	38.8	38.8	38.9	39.7
Urban population growth (%)	5.47	3.2	3.05	3.2
Gabon				
Total population	1,087,327	1,235,274	1,370,729	1,505,463
Population growth (%)	2.95	2.33	1.96	1.87
Urban population (% of total)	75.4	80.1	83.6	86
Urban population growth (%)	4.64	3.51	2.8	2.43

Source: Authors, from World Development Indicators database, World Bank (http://databank.worldbank.org/ddp/home.do; accessed March 2012).

Figure 2.2 Urban Population in the Congo Basin Countries, 1995–2010
% of the total population

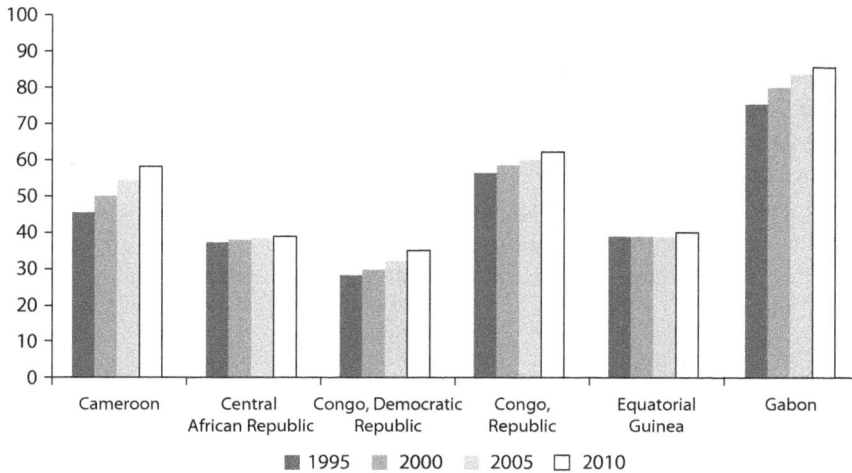

Source: Authors, from World Development Indicators database, World Bank (http://databank.worldbank.org/ddp/home.do; accessed March 2012).

100,000 inhabitants (cf. territories close to large urban centers). In rural areas, Zhang et al. also showed—through a GIS-based assessment on the vulnerability and future extent of the tropical forests of Congo Basin—that the annual clearance of the dense forest is significantly linked to rural population density. It also highlighted that there was a positive relationship between the dense forest degraded during the 1980s–1990s and the degraded forest area in the 1980s (Zhang et al. 2006). The transition zones between rainforest and savanna, where populations are usually much greater and can reach up to 150 inhabitants per square kilometer, also usually have considerable deforestation or forest degradation rates.

A Modeling Approach to Understand the Impact of Global Trends on the Congo Basin

The nature and amplitude of deforestation are likely to change significantly in the Congo Basin in the next two decades. Compared with other tropical forest blocks, deforestation and forest degradation have been globally low. They have been traditionally and dominantly caused by shifting cultivation and woodfuel collection in Central Africa; however, there are signs that the Basin forest is under increasing pressure and that deforestation is likely to soon magnify through the combined effects of the amplification of the existing drivers and the emergence of new ones.

- Current drivers of deforestation (internal) are expected to amplify. Demographic factors (population growth as well as rural/urban profile) are determinant causes of deforestation and forest degradation in the Congo Basin (Zhang et al. 2006). If existing rates of demographic growth remain constant, then the population of the Basin will double by 2035 to 2040. In most Basin countries, the population is still largely involved in subsistence farming and predominantly relies on woodfuel for domestic energy.

- New drivers, external by nature, are now emerging in the context of a more and more globalized economy. Congo Basin countries are poorly connected to the globalized economy and, accordingly, the drivers of deforestation have so far mainly been endogenous (essentially population driven). However, signs suggest that the Basin may no longer be immune to global demand for commodities—directly or indirectly—with increasing pressure from a variety of forces, including oil and mineral extraction, road development, agribusiness, and biofuels.

A modeling approach has been elaborated to investigate the effect of the predicted main future drivers of deforestation in the Congo Basin, both internal and external, on land use change and on resulting greenhouse gas (GHG) emissions by 2030. In fact, the high forest cover and low deforestation (HFLD) profile of the Basin countries justified the use of a prospective analysis to forecast deforestation, as historical trends were considered inadequate to properly capture the future nature and magnitude of drivers of deforestation. Accordingly, a macroeconomic approach, based on the GLOBIOM model (Global Biomass Optimization Model), was taken in order to sufficiently take into account global parameters.

GLOBIOM is a partial equilibrium model and is an economic model that incorporates only some sectors of the economy. Like all models, GLOBIOM simplifies a complex reality by highlighting some variables and causal relations that explain land use change based on a set of assumptions about an agent's behavior and a market's functioning (see box 2.1). GLOBIOM only includes the main sectors involved in land use, that is, agriculture, forestry, and bioenergy. It is an optimization model that searches for the highest possible levels of production and consumption, given the resource, technological, and political constraints in the economy

---

Box 2.1 More on the GLOBIOM Model: Underlying Assumptions

Assumptions: GLOBIOM relies mainly on neoclassical assumptions. Agents are rational: consumers want to maximize their utility and producers want to maximize their profits. The markets are perfectly competitive, with no entry and no exit costs and homogeneous goods, which imply that agents have no market power and that at the equilibrium the profits are equal to zero. The equilibrium prices ensure demand equals supply. Agents have a perfect knowledge—that is, no uncertainty is taken into account. We assume buyers are distinct from sellers so that consumption and production decisions are taken separately. Moreover, markets are defined at the regional level, meaning consumers are assumed to pay the same price across the whole region; however, selling prices could vary across the region because production costs and internal transportation costs are defined at the pixel level.

(McCarl and Spreen 1980). The demand in the GLOBIOM model is exogenously driven—that is, some projections computed by other teams of experts on population growth, gross domestic product (GDP) growth, bioenergy use, and structure of food consumption are used to define the consumption starting point in each period in each region. Then, the optimization procedure ensures that the spatial production allocation minimizes the resources, technology, processing, and trade costs. Final equilibrium quantities result from an iterative procedure between supply and demand, where prices finally converge to a unique market price. Box 2.1, as well as the appendix, provide a detailed description of the GLOBIOM model.

GLOBIOM is designed for the analysis of land use changes around the world.[1] The biophysical processes modeled (agricultural and forest production) rely on a spatially explicit data set that includes soil, climate/weather, topography, land cover/use,[2] and crop management factors. Harvesting potentials in cropland are computed with the Environmental Policy Integrated Climate (EPIC) model (Williams 1995) that determines crop yields and input requirements based on relationships among soil types, climate, hydrology, and so on. Timber-sustainable harvesting potential in managed forests is computed from the G4M model's forest-growth equations. The GLOBIOM model draws on extensive databases for initial calibration of the model in the base year, technical parameters, and future projections. In order to reproduce the observed quantities for the reference year (2000), the GLOBIOM model is calibrated by employing the Positive Mathematical Programming (Howitt 1995) that consists of using the duals on the calibration constraints to adjust the production cost. This process is supposed to correct the model's problems of specification and the omission of other unobservable constraints that face production. It is used to calibrate the crop, sawn wood, wood pulp, and animal calories production.

GLOBIOM is a global simulation model that divides the world into 28 regions. One such region considered under the GLOBIOM model is the Congo Basin (the six highly forested countries covered by the study). It is important to look at the rest of the world when studying land use change in a region because local shocks affect international markets and vice versa. Moreover, there are important leakage effects. Bilateral trade flows are endogenously computed between each pair of regions, depending on the domestic production cost and the trading costs (tariff and transportation costs).

Based on an adaptation of GLOBIOM, the CongoBIOM model has been elaborated.[3] The Congo Basin region was specifically created within the GLOBIOM model, and additional detail and resolution for the Basin countries were included. Land-based activities and land use changes have been modeled at the simulation-unit level, which varies in size between 10×10 kilometers and 50×50 kilometers. Internal transportation costs have been computed based on the existing and planned infrastructure network; protected areas and forest concessions have been delineated, and available national statistics have been collected to inform the model (IIASA 2011; Mosnier et al. 2012). The calibration of the CongoBIOM model was done on the data collected in the various six countries by a team of international and national experts.

The CongoBIOM was used to assess the impacts of a series of "policy shocks" identified by Congo Basin country representatives. The methodological approach was first to investigate what could be the reference level of emissions from deforestation in the Congo Basin without further measures to prevent or limit deforestation. Complementary scenarios were tested in addition to the baseline with different assumptions about global meat and biofuel demand, internal transportation costs, and crop yield growth (table 2.2). The selection of the policy shocks was based on a literature review and was validated during two regional workshops with local experts. Policy shocks were chosen to describe impacts from both internal and external drivers: external (S1): increase in international demand for meat; (S2): increase in international demand for biofuels; (S3): improved transport infrastructure; (S4): decrease in woodfuel consumption; and (S5): improved agricultural technologies. Table 2.2 describes the scenarios used under the modeling exercise in the Congo Basin (as well as the main results). The objectives were (1) to highlight the mechanisms through which deforestation could occur in the Basin (driven by both internal and external drivers); and (2) to test the sensitivity of deforested area and GHG emissions from deforestation with respect to different drivers.

Data availability and quality have been major challenges for the modeling approach. Spatially explicit input parameters are mainly related to resource availability, production costs, and production potentials. Crop-harvested areas and forest carbon stocks have been allocated at the pixel level by downscaling methodologies also subject to errors. Uncertainty about land cover is especially prevalent in the Congo Basin due to the permanence of clouds and the limited number of images in the past. Despite the significant effort to enhance both availability and quality of the data used in the model (through a data collection campaign in all six countries), limitations persisted and consequently a decision was made that the modeling exercise would be primarily used to strengthen the understanding of deforestation dynamics and causal chains (internal/external drivers) in the Basin. The quantitative outputs of the model presented in box 2.2 should be taken with extreme caution and used, rather, as a comparative basis between the different scenarios. Validation of these input data would require additional statistics at a finer resolution level and would ideally be available for several years.

The CongoBIOM model provides key insights into the interconnectivity of macroeconomic developments and the exposure of the Congo Basin forests to exogenous shocks and threats. The results of the CongoBIOM model, wherever available, will be presented throughout the following section of this chapter. The main outcomes of the exercise follow and will be further presented in later sections.

- The model confirms that pressures on Congo Basin forests could increase in the coming decades. It highlights that so far the Basin has been quite isolated from international markets due to important internal handicaps. This separation is one reason for the historically low rate of deforestation in the

Table 2.2 Policy Shocks Tested with CongoBIOM and Main Results

Scenarios	Description	Main results
Baseline	Business as usual using standard projections of main model drivers.	Deforestation rate close to the historical rate of deforestation over 2020 to 2030 (0.4 Mha per year). Productivity gains avoid about 7 Mha of cropland expansion (the equivalent of the projected cropland expansion).
S1: Meat	Business as usual with a higher global meat demand. In the scenario, the demand of animal calories increases by 15% compared to Food and Agriculture Organization of the United Nations (FAO) projection in 2030.	The Congo Basin countries remain marginal in meat production. The average deforested area over the 2020–30 period still increases by 20% in the Congo compared to the base Basin. As the global price for meat and animal food increases, food and feed imports are reduced and local production increases—leading to deforestation.
S2: Biofuels	Business as usual with a higher global first-generation biofuel demand. The scenario on the biofuel consists of doubling the demand for biofuels of first generation compared to the initial projection of the POLES model in 2030.	The Congo Basin countries remain marginal in global biofuels and feedstock production. The average deforested area over the 2020–30 period still increases by 36% in the Congo Basin compared to the base. As the global price for oil palm and agriculture products increase, food imports are reduced and local production of oil palm and food increases—leading to deforestation.
S3: Infrastructure	Business as usual with planned transportation infrastructure included. Return of political stability, good governance, and new projects induced a multiplication of projects to repair existing transport systems and contribute to a new transportation system. The model has included all the projects for which the funding is certain.	Calorie intake per capita increases by 3% compared to the base scenario. The Congo Basin improves its agricultural trade balance with an increase in exports and a reduction in food imports. Total deforested area becomes three times as large (+234%) and emissions from deforestation escalate to more than four times as large.
S4: Woodfuel	Business as usual with a decrease in woodfuel consumption per inhabitant from 1 m³ to 0.8 m³ per year.	Within the 0.4 Mha deforested per year on the baseline, woodfuel counts for 30%. A 20% decrease in woodfuel consumption induces therefore a 6% decrease in total deforestation compared with the business-as-usual scenario.
S5: Technological change—Increase in agriculture productivity	Business as usual with increased crop productivity. The model assumes that this increase is proportional across all management systems and does not involve higher production costs for farmers (modeling, for example, agricultural mechanization or subsidies of better seeds). The yields are doubled for food crops and increased by 25% for cash crops.	Calorie intake per capita increases by 30% and imports are reduced. Increase in emissions from deforestation by 51% over the 2020–30 period because consumption increases faster than that of crop productivity.

Source: International Institute for Applied Systems Analysis [IIASA] 2011.
Note: Mha = Millions of hectares.

Box 2.2 Results of the CongoBIOM Model

The annual average deforested area varies between 0.4 to 1.3 million hectares over 2020 to 2030 across the different scenarios: The scenarios with the highest negative impact on forest cover in the Congo Basin are the improvement in transportation infrastructure and the technological change—that is, the internal drivers of deforestation (see figure B2.2.1).

It suggests that future deforestation levels depend mainly on the domestic policies that will be implemented in the Basin.

Figure B2.2.1 Results of the Policy Shocks in Terms of Areas Annually Deforested under the Different Scenarios, 2020–30

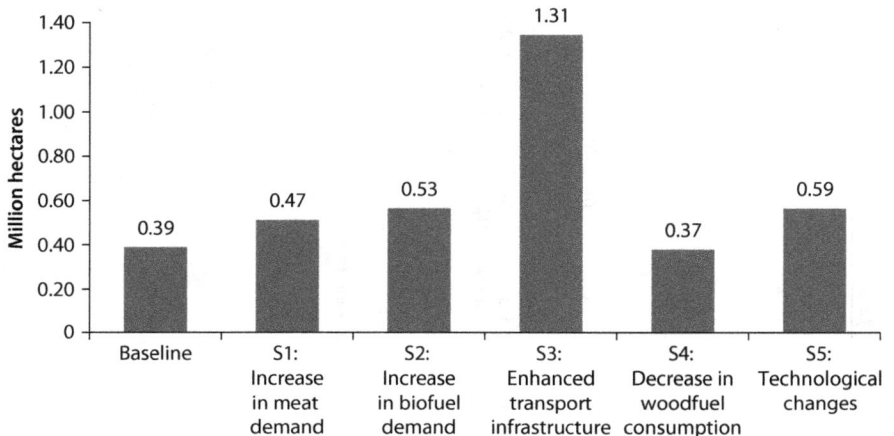

Source: IIASA 2011.

region—as compared to that of other tropical regions that have become major exporters of agricultural products during the last decades. However, as the Congo Basin countries become more exposed to international markets and the population grows, this situation is most likely to change. Deforestation in the Basin is likely to increase in the near future.

- Future deforestation in the Congo Basin depends critically on national strategies to be implemented in the next years, and on local investments. The scenarios with the highest negative impact on forest cover in the Basin are the improvement in transportation infrastructure (S3) and the technological change (S5) that is, the internal drivers of deforestation.

- International drivers (S1 and S2) are likely to trigger additional deforestation in the Congo Basin; however, their impact is likely to be limited to a higher substitution of imports with local production. CongoBIOM indicates that the Basin would be much less affected than the Southeast Asia tropical forests and the Amazonian tropical forests under the S2 scenario. Moreover, the impacts would be rather indirect. In fact, the costs of doing business in the region continue to be very high; the risks associated with poor governance represent a significant financial burden to investors, while unclear legislation on land tenure is not conducive for agroindustrial investments. So far, other tropical regions provide investors with a more conducive investment climate (for example, Southeast Asia and Latin America). In a nutshell, deforestation in the Congo Basin is likely to increase when (1) import prices increase more than domestic prices so that local consumers shift a part of their consumption from imports to local production; (2) lower domestic prices stimulate the demand for local products; and (3) the Basin improves its competitiveness on international markets and increases its exports.

However, as the Congo Basin countries prepare the REDD+ strategy, one should keep in mind the limitations of the CongoBIOM model, particularly as a decision-making tool on REDD+ (see box 2.3).

Box 2.3 Limitations of the CongoBIOM Model and Way Forward in the REDD+ Context

- The limited availability and poor quality of data did not allow for an analysis of the potential trade-offs between the different scenarios and REDD+. In addition, there is no clear signal on the level of benefits that could be yielded from the REDD+ mechanism. As such, the assessment of such trade-offs has been considered premature. Additional analytical work should be conducted to further assess and quantify trade-offs between economic growth and forest preservation.

- The CongoBIOM model is a spatial model, but assumptions and the variant calibration are made at the regional, not country, level. Building a national model would require a dramatic increase in data since intrabasin trades should be identified. Also, variables that were homogenized at the regional level (such as prices) would need to be defined for each country. Downscaling the model to the country level could be a future development, depending on countries' interest.

- At this point of the process, the CongoBIOM model cannot be used as a precise reference scenario. As the objective of the model was to describe the main trends, it provides an order of magnitude of the deforestation. Precise figures for reference levels would require more detailed data on the countries' planned development and investments.

- Using CongoBIOM as a decision-support tool would require several iterations of the model calculation under different assumptions. Therefore, much data would be required and the model developed further before it could be fully used as policy adviser. Working with academic institutions in each country would then help to disseminate knowledge and to build capacity.

Agriculture Sector

A Vital yet Neglected Sector

Most of the rural households in the Congo Basin rely primarily on agricultural activities for their livelihoods. Agriculture remains by far the largest employer in the region. In Cameroon, the Democratic Republic of Congo, the Central African Republic, and Equatorial Guinea more than half of the economically active population is still engaged in agricultural activities, although there is a declining trend in the share of employment in this sector across all countries (figure 2.3).

Agriculture remains a chief contributor to GDP, especially in the Central African Republic, the Democratic Republic of Congo, and Cameroon. Agriculture's contribution to GDP remains high at around 40 to 50 percent in the Central African Republic and the Democratic Republic of Congo (figure 2.4). In the latter, an unstable political context has led to a strong variability of agriculture contribution to GDP over the past two decades (see Note under figure 2.4). Agriculture's contribution to GDP is logically much lower in the other four oil-producing countries, although it is still around 20 percent for Cameroon, which has a much stronger agricultural basis than do Equatorial Guinea, Gabon, and the Republic of Congo. The contribution of agriculture to GDP in Equatorial Guinea dropped dramatically during the mid-1990s due to the sharp increase in oil revenues (total GDP increased by a factor of 60).

The agriculture sector has so far been neglected and underfunded for much of the last few decades. Public expenditure in agriculture in all six countries is lagging far behind the 10 percent of total national budget targeted by the CAADP[4] initiative, affecting primarily extension services, basic infrastructure (feeder roads), and research and development (R&D; see table 2.3 and box 2.4). The natural resources curse (Collier 2007), also known as the paradox of plenty, is arguably a major reason why the agricultural sector has received so little

Figure 2.3 Congo Basin Countries' Share of Economically Active Population in Agriculture

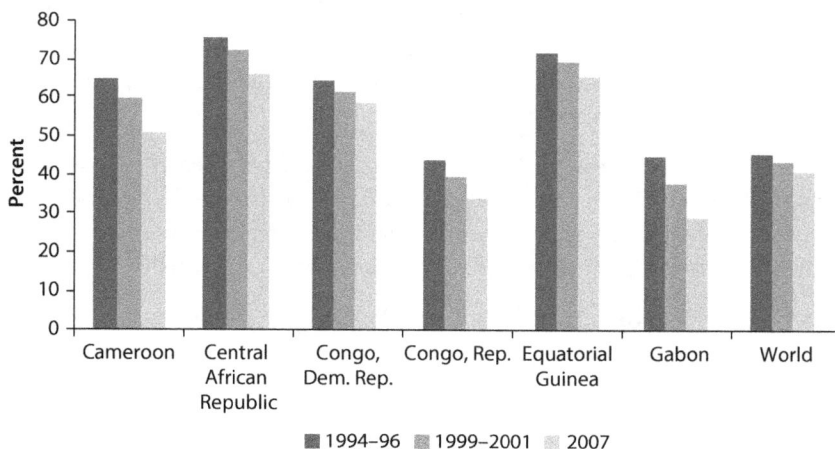

■ 1994–96 ▨ 1999–2001 ▦ 2007

Source: FAO 2009b.

Figure 2.4 Congo Basin Countries' Evolution of Agriculture's Contribution to GDP, 1988–2008

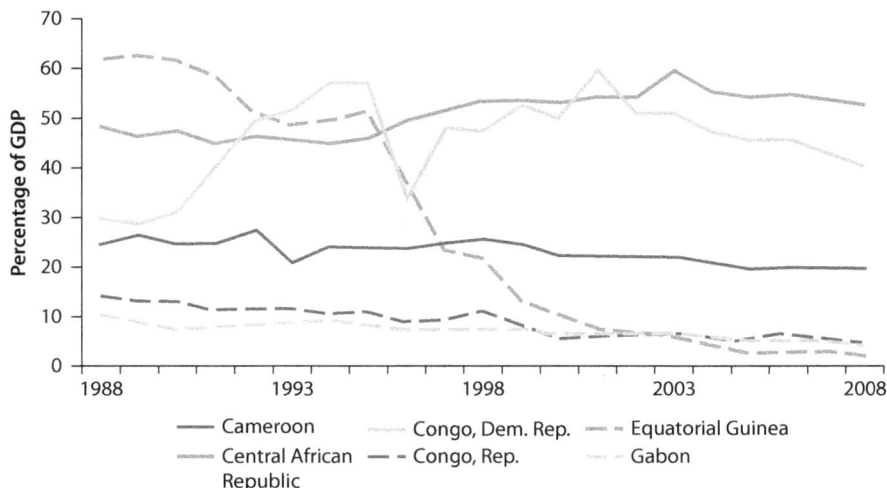

Source: Authors, prepared with data from World Development Indicators, World Bank (http://databank.worldbank.org/ddp/home.do; accessed March 2012).
Note: In the late 1980s to early 1990s, the sharp decline of the Democratic Republic of Congo's economy resulted in agriculture representing an increased share of the gross domestic product (GDP). Agriculture was then severely disrupted by the 1996 civil war. The recovery of agriculture's contribution to GDP after the war was more a reflection of the poor overall economic situation rather than an indication that agricultural output was growing over that period.

Table 2.3 Congo Basin Countries' Share of Agricultural Expenditure in National Budget
percent

	%	Year reported
Cameroon	4.5	2006
Central African Republic	2.5	—
Congo, Dem. Rep.	1.8	2005
Congo, Rep.	0.9	2006
Equatorial Guinea	—	—
Gabon	0.8	2004

Source: ReSAKSS (http://www.resakss.org; accessed February 2012); no data available for Equatorial Guinea.
Note: Trying to determine total agricultural research and development (R&D) public sector spending in the Congo Basin countries based on the database set up by the International Food and Policy Research Institute (IFPRI) on Agricultural Science and Technology Indicators (ASTI) proves difficult because most of the Basin countries, unlike those in western or eastern Africa, do not report data. The only data available are for Gabon (2001) and the Republic of Congo (2001), $3.8 million and $4.7 million, respectively, in 2005 U.S. dollars, which are among the lowest figures of R&D public budgets in Sub-Saharan Africa. It is also known that the Central African Republic, the Democratic Republic of Congo, and Equatorial Guinea spend very little on agricultural research. Only Cameroon in Central Africa has a performing national agricultural-research institute, *IRAD (Institut de Recherche Agricole pour le Développement)*, with about 200 researchers in 10 research stations but also with minimal operating funds. — = not available.

interest over the past decades. Because Congo Basin countries are richly endowed with natural resources, particularly such nonrenewable ones as oil and minerals, they tend to neglect their agriculture and import most of their food. In addition to policy makers' disinterest, the boom in extractive industries and associated revenues generates discriminatory conditions against other productive economic

---

Box 2.4 Agriculture: Evolution in Public Policies in the Congo Basin

Until the late 1980s, as in almost all Sub-Saharan African countries, the negative impact of public resources scarcity was aggravated by fiscal and trade policies that strongly discriminated against agriculture and discouraged investments from both local farmers and foreign operators.[a] With the exception of Cameroon, where some supportive policies were implemented, the other Congo Basin countries did not set the basic conditions to realize their agricultural potential.

In the 1990s, all countries went through a structural adjustment process with associated dramatic cuts in public expenditures in order to reduce the substantial external and internal deficits of their economies. Net taxation of agriculture decreased, but the agricultural sector was one of the most impacted by budgetary restrictions: fertilizer and pesticide subsidies (ranging from 60 to 100 percent in Cameroon) were removed, extension services drastically reduced, rural infrastructure neglected, and research and development almost abandoned. At the same time, major reforms occurred in the export-oriented agricultural sector (such as coffee and cocoa), with the state disengaging and liquidating the national marketing boards for these crops.

More recently, the lukewarm response of the Basin countries to the continent-wide New Partnership for Africa's Development (NEPAD) initiative in favor of agriculture, the Comprehensive Africa Agriculture Development Program (CAADP), which targets an annual 6 percent agricultural growth through, in particular, greater government support to the sector, suggests that these countries' governments still do not view agriculture as a critical cornerstone to development, food security, and poverty alleviation. Although 22 countries have already signed their CAADP compacts and made substantial progress toward their goals, none of the Congo Basin countries have followed suit.

a. It is estimated that in the 1980s, net taxation of the agricultural sector in Sub-Saharan Africa through overvalued exchange rates, controlled input and output prices, export taxes, and so on, averaged 29 percent and stood at 46 percent for exportables (World Bank 2009).

---

sectors, including a decline in competitiveness caused by the appreciation of the real exchange rate as resources enter the economy in large amounts. However, there have been recent signals of heightened interest in the agriculture sector in most of the Congo Basin countries, as evidenced by the medium- to long-term strategies for development prepared by these countries[5] that include agriculture as one of the economic pillars for development and growth. Interestingly, these strategies cover both commercial and subsistence agriculture as complementary segments of the sector.

Agricultural production is still largely ruled by traditional subsistence systems. The agricultural sector in the Congo Basin is dominated by smallholders[6] who cultivate a maximum of 2 to 3 hectares of traditional crops on a 2-year cultivation and 7- to 10-year fallow pattern.[7] Maize, groundnuts, taro, yams, cassava, and plantains are grown mostly for their own consumption, with possibly a surplus sold in the local market.[8] Some traditional slash-and-burn smallholder farms

What Will Drive Deforestation in the Congo Basin? A Multisectoral Analysis

71

plant cocoa, coffee, and oil palm. Coffee and cocoa are primarily produced on areas of 0.5 to 3 hectares[9] (Tollens 2010).

There are also a few large commercial enterprises—usually owned by multinational companies active in the region—in particular in palm oil and rubber production (and bananas in the case of Cameroon). Large-scale plantations can be seen as enclaves of the modern sector within the traditional sector, with very little if any relations between them. Oil palm is cultivated in both smallholder plantations (100 percent of production in the Central African Republic, Equatorial Guinea, and the Republic of Congo; 85 percent in the Democratic Republic of Congo) and in large-scale operations by multinational companies (Gabon, Cameroon). The Congo Basin has not witnessed yet, however, the expansion of large-scale plantations as other tropical regions have experienced. The Basin countries have—so far—been spared the phenomenon of large-scale land acquisition and conversion for agriculture and biofuel projects that has been observed in other regions of the world (Southeast Asia, Amazonia). The few current operators who exist in Cameroon, Gabon, and the Democratic Republic of Congo report that they do not plan to invest in new plantations; they intend to extend existing concessions and rehabilitate old or abandoned ones (Tollens 2010; see box 2.5 and please also refer to box 2.10).

Box 2.5 Recent Trends in Large-Scale Land Acquisition and Their Effects on the Democratic Republic of Congo

As a result of the 2007–08 spike in food prices and relatively meager returns from other assets due to the financial crisis, a number of investors recently turned their attention to agricultural production in developing countries. According to press reports, the ensuing wave of interest in land acquisition amounted to some 57 million hectares in less than a year, largely outweighing the 1.9 million hectares average annual rate of cropland expansion between 1990 and 2007. Countries with relatively large tracts of nonforested, uncultivated land with agricultural potential, but also countries with poorer records of formally recognized land rights, attracted more investor interest, especially land-abundant countries in Sub-Saharan Africa that, according to media reports, represented more than two-thirds of the expressed interest (40 million hectares).

In 2010–11, the World Bank implemented a survey of recent investments in the Democratic Republic of Congo to determine how much land had been demanded by investors and for what purpose, as well as the extent to which any land granted to investors had been put to use (Deininger et al. 2011). The results indicate that demand for farmland in the Democratic Republic of Congo concentrated in the savanna regions in terms of the number of projects but in the forest zone in terms of areas: of the 42 projects that applied for land rights greater than 500 hectares between 2004 and 2009, only 4 were located in heavily forested regions (Bandundu, Equateur, and Orientale Provinces). They focused on biofuels,

box continues next page

Box 2.5 Recent Trends in Large-Scale Land Acquisition and Their Effects on the Democratic Republic of Congo *(continued)*

industrial crops (rubber), and reforestation, but these four projects account for roughly 420,000 hectares—that is, nearly 76 percent of the total area requested by investors. If these projects become operational, they could have significant impacts on forests in the Congo Basin. However, the large gap between expressed demand for land and actual investment implementation in the Democratic Republic of Congo suggests that the impact of large-scale agricultural investments in Basin countries is likely to take time to materialize, unless endogenous variables (business environment, infrastructure) and exogenous ones (food and biofuel demand and supply) change significantly in the short term.

Source: Deininger et al. 2011.

Productivity for most commodities grown in the Congo Basin is very low compared to that of other tropical countries. The reliance on mostly vegetatively propagated crops considerably slows down the dissemination of improved varieties. The use of fertilizers and pesticides is also among the lowest in Africa with, for fertilizers, an average of fewer than 2 kilograms per hectare, with the exception of Cameroon and Gabon, where 7 to 10 kilograms per hectare are reported to be applied.[10] As a result, yields for most commodities grown, either staples or cash crops, are particularly low in the Basin (as shown in the series of graphs in figure 2.5, where Congo Basin countries are represented in orange and other countries from tropical regions are shown in blue). The only exception is palm oil production in Cameroon, with observed yields among the highest in the world, comparable to leading countries.

Agricultural trade balances have deteriorated. Except in the Central African Republic, where agricultural import and export values have changed little around the equilibrium over the past 15 years, the agricultural trade balance has severely deteriorated in all countries of the Congo Basin (figure 2.6). All Basin countries except the Central African Republic are net importers of food[11] (table 2.4). Recent statistics from Food and Agriculture Organization of the United Nations (FAO) show that food commodities imports are rising rapidly and that these countries rely more and more on imports to fill their basic food needs. A large proportion of these rapidly increasing imports reflect urban-based shifts in consumption patterns toward more cereals (wheat and rice); fewer roots, tubers, and coarse grains; more animal proteins (chicken meat and eggs); and more readily prepared convenience foods.

A Sizable Potential for Agricultural Development in the Congo Basin

The potential for agricultural development in the Congo Basin is significant. The region is among the areas with the greatest potential in the world for both expanding cultivation and increasing existing yields; however, it remains to be seen whether and to which extent this potential manifests over the course of

Figure 2.5 Yields in Congo Basin Countries Compared with the Yields Obtained in Major Producer Countries, 2009

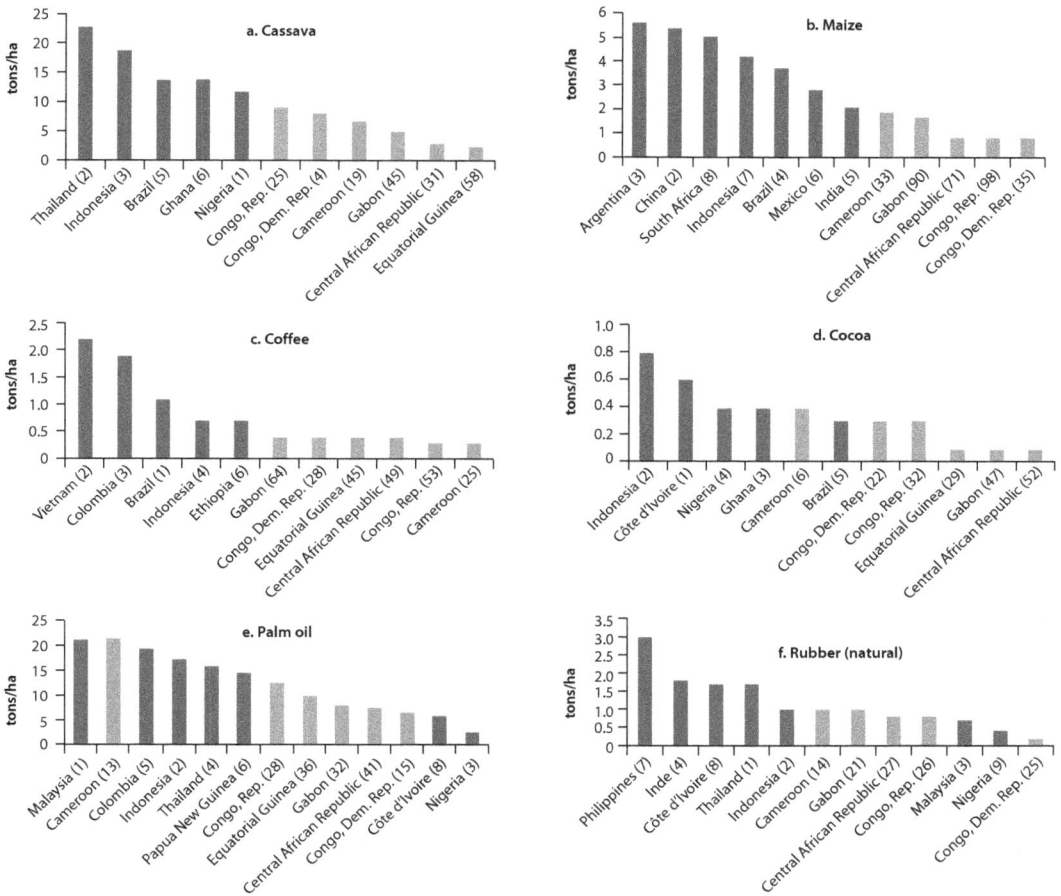

Source: FAOSTAT. 2011. http://faostat.fao.org/, FAO, Rome (accessed December 2011).
Note: Numbers in parentheses indicate the country's world rank in terms of the production of that commodity in 2008.

the next decades. Market forces, driven by both internal (domestic and regional markets) and external drivers (growing international demand for food and energy), suggest that agriculture will, in the medium and long term, expand. The section below presents the major factors likely to influence the agricultural development in the Congo Basin.

Vibrant Markets: Domestic, Regional, and International
The fast-rising urban population will drive an increase in internal demand for food. With the rapid growth of urban centers in the region, the demand for staple foods and convenience foods such as bread, rice, eggs, chicken, fish, and palm oil is growing. Currently, all six Congo Basin countries are net importers of these commodities. Import substitution could support agricultural growth in

Figure 2.6 Congo Basin Countries' Evolution of the Agricultural Trade Balance, 1994–2007

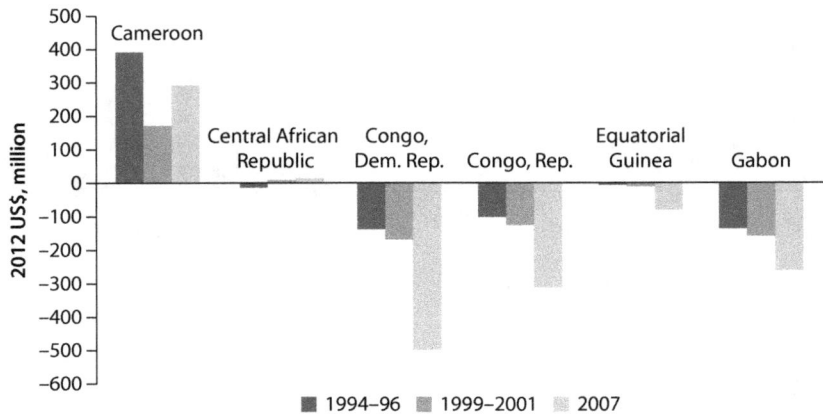

■ 1994–96   ■ 1999–2001   ▨ 2007

Table 2.4 Congo Basin Countries' Net Food Trade, 2006

Country	% of GDP
Cameroon	−0.7
Central African Republic	−0.5
Congo, Dem. Rep.	−4.9
Congo, Rep.	−2.6
Equatorial Guinea	—
Gabon	−2.3

these countries: products could be locally grown while some new products could advantageously replace imported ones (for instance, cassava-based flour has the potential to replace imported wheat flour, as has happened in West Africa).

The regional market is still to be unlocked. Agriculture markets in Central Africa are largely segmented. Deteriorated infrastructure and high transaction costs hamper the development of agricultural trade and exchanges not only at the national level but more broadly at the regional level. Unlocking these markets and the exchanges at the regional level could provide a boost for agriculture in the subregion. Presently, most of the trans-boundary fluxes, though quite active, are informal; they cover all kind of products (staple products and plantation crops). Formalizing these fluxes, through regional trade agreements and regional integration, could support agricultural growth in Central Africa.

Prospects for exports at the international level are promising. There are positive price projections for most of the commercial crops grown in the subregion.

What Will Drive Deforestation in the Congo Basin? A Multisectoral Analysis

75

Palm oil is now the most predominantly used oil at the international level, and the evolution of biofuel demand could amplify the demand for oil palm plantations. Rubber, though largely affected by the financial crisis and the subsequent car manufacturing crisis, is showing good trends with a rising demand from emerging markets in India and China. Cocoa is the only agricultural commodity that has not been affected by the contraction of the markets during the financial crisis and showed strong performances that are likely to be maintained. The coffee price is much more volatile but could still present opportunities.

International demand for food and energy is rising. Experts estimate that the foreseen increase of 40 percent in the world population by 2050, combined with a rise in average food consumption, will require a 70 percent increase in agricultural production (100 percent in developing countries; Bruinsma 2009). In this context, FAO projections suggest that, although less strong than in the past, yield increases (along with increased cropping intensity) will still account for 90 percent (80 percent in developing countries) of production growth, with the remainder coming from land expansion. That would translate into 47 million hectares of land to be brought into production globally over the 2010–30 period, with a decrease of 27 million hectares in developed and transitional countries, and an increase of 74 million hectares in developing economies. Demand for feedstock (wheat, maize, sugarcane, oil seeds), which are not included in the above projections, will also be a major factor driving world agriculture evolution, with land conversion for biofuels by 2030 estimated to range between 18 and 44 million hectares.

Land Availability/Suitability/Accessibility in the Congo Basin

Land suitability: The Basin countries rank just behind Latin America in terms of suitability for major export crops (such as soybean, sugar cane, and oil palm). In the Democratic Republic of Congo, a 2007 agricultural suitability mapping exercise estimated that about 60 percent of the dense humid forest was fit for the production of palm oil (approximately 47 million hectares [ha]; Stickler et al. 2007).

Land availability: A recent study commissioned by the World Bank modeled the potential worldwide availability of land for rain-fed crop production (Deininger et al. 2011). Altogether, Congo Basin countries represent about 40 percent of the uncultivated, unprotected, low-population density land suitable for cultivation in Sub-Saharan Africa and 12 percent of the land available at the world level[12] (table 2.5). The ratio of suitable to cultivated land, particularly high in the Basin countries, illustrates the great potential for investments in land expansion (see the section on Land Availability: Forested versus Nonforested).

Land accessibility: Poor road infrastructure in the Congo Basin has been a major and long-lasting obstacle in the transition to a more performing agriculture mainly because potential areas available and suitable to agricultural production are too remote from markets. An International Institute for Applied Systems Analysis (IIASA) model identifies potentially suitable and accessible land by computing production cost estimates in order to arrive at

Table 2.5 Potential Land Availability by Country

million ha

				Suitable noncropped, unprotected area density < 25 people/km²	
	Total area	*Forest area*	*Cultivated area*	*Forest*	*Nonforest*
Sub-Saharan Africa	2,408.2	509.4	210.1	163.4	201.5
Congo, Dem. Rep.	232.8	147.9	14.7	75.8	22.5
Sudan	249.9	9.9	16.3	3.9	46.0
Zambia	75.1	30.7	4.6	13.3	13.0
Mozambique	78.4	24.4	5.7	8.2	16.3
Angola	124.3	57.9	2.9	11.5	9.7
Madagascar	58.7	12.7	3.5	2.4	16.2
Congo, Rep.	34.1	23.1	0.5	12.4	3.5
Chad	127.1	2.3	7.7	0.7	14.8
Cameroon	46.5	23.6	6.8	9.0	4.7
Tanzania	93.8	29.4	9.2	4.0	8.7
Central African Republic	62.0	23.5	1.9	4.4	7.9
Gabon	26.3	21.6	0.4	6.5	1.0
Latin America and the Caribbean	2,032.4	934.0	162.3	290.6	123.3
Eastern Europe and Central Asia	2,469.5	885.5	251.8	140.0	52.4
East and South Asia	1,932.9	493.8	445.0	46.3	14.3
Middle East and North Africa	1,166.1	18.3	74.2	0.2	3.0
Rest of the world	3,319.0	863.2	358.9	134.7	51.0
World Total	13,333.1	3,706.5	1,503.4	775.2	445.6

Source: Deininger et al. 2011.
Note: In Sub-Saharan Africa, only countries that have more noncropped, unprotected suitable land (forest or nonforest) than Gabon are detailed here. ha = hectare; km² = square kilometer.

the net profits rather than the revenues: the possibly suitable land was further classified based on the travel time to the next significant market, defined as a city of at least 50,000 inhabitants, with a cutoff of six hours to market. As shown in table 2.6 below, Latin America has a clear advantage infrastructure-wise, with more than 75 percent of its nonforested suitable land at less than six hours from a market town, against less than 50 percent in Sub-Saharan Africa. Consequently, despite Latin America having about 40 percent less land available than does Sub-Saharan Africa, the regions contain roughly the same amount of nonforested suitable land (about 94 million ha) when the access-to-market criterion is factored. The situation is even worse in the Basin countries: in the Democratic Republic of Congo, it is estimated that only 33 percent (7.6 out of 22.5 million ha) of the nonforested suitable land is at less than six hours from a major market; that proportion is as low as 16 percent in the Central African Republic (1.3 out of 7.9 million ha).

Table 2.6 Land Accessibility: Potential Supply of Uncultivated, Nonforested, Low Population Density Land

	Total area (million ha)	Area < 6 hr to market (million ha)	% Area < 6 hr to market
Sub-Saharan Africa	201.5	94.9	47.1
Latin America and the Caribbean	123.3	94.0	76.2
Eastern Europe and Central Asia	52.4	43.7	83.4
East and South Asia	14.3	3.3	23.1
Middle East and North Africa	3.0	2.6	86.7
Rest of the world	51.0	24.6	48.2
Total	445.6	263.1	59.0

Source: Deininger et al. 2011.
Note: Low population density ≤ 25 persons/km². ha = hectare; hr = hour.

Potential to Increase Productivity

The Congo Basin is among the areas with the greatest potential in the world for increasing existing yields. The methodology developed by IIASA, using high-resolution agroecological zoning, was employed by the World Bank to predict land suitability, potential yields, and gross value of output for five key crops: wheat (not relevant in the case of Basin countries), maize, oil palm, soybean, and sugarcane. The model highlights the Basin as one of the areas with the greatest maximum potential value of output in the world for these crops.

Water Resources Availability

Water resources in the Congo Basin are expected to be unconstrained, thus providing strong support for agricultural development. Many parts of the world, particularly developing countries, are expected to experience water scarcity and stresses in the future. Water deficiency and competition with other uses in many regions (such as China, South Asia, the Middle East, and north Africa) will have profound impacts on agricultural production, including possible changes in cropping patterns, reduced yields, greater frequency of extreme weather events (resulting in higher variability of output), and the need in certain areas to invest in water-storage infrastructure to capture more concentrated rainfall and minimize associated soil erosion. In this changing climate context, Basin countries show a profile where water resources are likely to increase or at least be maintained. In addition, compared with some neighboring countries, Basin countries have been globally spared by natural disasters related to weather extremes. This resilience to climate change may potentially provide the Congo Basin countries with a distinct agricultural advantage at the global level.

Impacts on Forests: Current and Future

Future agricultural development may be at the expense of forests. The factors previously described suggest that the agriculture sector could take off during the next decades. Unlocking this potential may lead to greater pressures on forests.

In fact, over the period from 1980–2000, more than 55 percent of the new agricultural land came at the expense of intact forests and another 28 percent from degraded forests (Gibbs et al. 2010).

The CongoBIOM model has identified the potential impacts of specific changes, both internal (such as land productivity) and external (international demand for meat or palm oil) on Congo Basin forests. Although the model has clear limitations and the data collected in the Congo Basin to run the different scenarios are generally of poor quality, it allows for a better understanding qualitatively (rather than quantitatively) of the dynamics and causal chains that can impact the Congo Basin forests in a globalized context.

Land Availability: Forested versus Nonforested Lands

While the vast majority of suitable land (noncropped, nonprotected areas) currently lies under forests, the potential of nonforested suitable lands is also considerable in the Basin and represents more than the area currently under production in most of the countries: the mean ratio of cultivated area to nonforested suitable area for Basin countries is 0.61, ranging from 1.45 in Cameroon to 0.14 in the Republic of Congo—way below the same ratio at the world level (3.37). This means that the Basin could almost double its cultivated area without converting any forested areas (see table 2.7 and box 2.6).

Increase in Land Productivity: Will It Reduce or Exacerbate the Pressure on Forests?

Without accompanying policies and measures on land planning and monitoring, a rise in agricultural productivity could lead to more deforestation in the Congo Basin. Increase in land productivity is often seen as the most promising means to jointly achieve the food production and the mitigation challenges. In fact, it is assumed that producing more on the same area should result in avoiding the conversion of new lands into agricultural production and that

Table 2.7 Ratio of Current Cultivated Areas to Potential Nonforested Suitable Lands in the Congo Basin

	Total land area	Cultivated area (A)	Suitable noncropped, nonprotected area density < 25 people/km²		Ratio A/B
			Forest	Nonforest (B)	
Cameroon	46.5	6.8	9.0	4.7	1.45
Central African Republic	62.0	1.9	4.4	7.9	0.24
Congo, Dem. Rep.	232.8	14.7	75.8	22.5	0.65
Congo, Rep.	34.1	0.5	12.4	3.5	0.14
Gabon	26.3	0.4	6.5	1.0	0.40
Total Congo Basin (except Equatorial Guinea)	401.70	24.30	108.10	39.60	0.61

Source: Authors, calculated from Deininger et al. 2011.
Note: km² = square kilometer.

Box 2.6 Land Suitability on Nonforested Areas in Congo Basin Countries

The Democratic Republic of Congo has the greatest reserve of uncultivated, nonprotected, and low population density[a] land suitable for cultivation in Sub-Saharan Africa. That reserve is estimated at 98.3 million hectares, of which three-fourths are currently under forest, and represents nearly seven times the area presently cultivated in this country (more than 16 times if the Food and Agriculture Organization of the United Nations (FAO) figure for the Democratic Republic of Congo's currently cultivated land is used).[b] If only *nonforested* suitable land is considered, the Democratic Republic of Congo still ranks among the six countries with the largest amount of suitable (but uncultivated) land available in the world (Sudan, Brazil, the Russian Federation, Argentina, Australia, and the Democratic Republic of Congo, in that order) but comes second to Sudan in Sub-Saharan Africa. The Democratic Republic of Congo's *nonforested* suitable (uncultivated) land is estimated at more than 1.5 times its currently cultivated land (and almost four times its currently cultivated land if the FAO figure is used).

Cameroon is estimated to have a reserve of 13.6 million hectares, of which about two-thirds are presently under forest. This is about twice its area presently under cultivation and 70 percent if only *nonforested* suitable land is considered.

The Republic of Congo is estimated to have 15.8 million hectares available (of uncultivated, suitable land), of which also about three-fourths are currently under forest. This reserve represents more than 30 times its area presently cultivated and still seven times if only *nonforested* suitable land is considered.

The Central African Republic is estimated to have a reserve of 12.3 million hectares, approximately one-third under forest, which represents more than six times its area currently cultivated and still more than four times if only *nonforested* suitable land is considered.

Gabon is estimated to have 7.4 million hectares available, almost 90 percent of which is currently under forest, representing about 19 times its area presently under cultivation. If only *nonforested* suitable land is considered, the land potentially available in that country still amounts to 2.5 times its currently cultivated land.

Source: Deininger et al. 2011.
a. A threshold of 25 persons per square kilometer—that is, more than 20 hectares per household—was used, under which the authors of the IIASA study considered that voluntary land transfers that make all stakeholders better off can easily yield agreement.
b. FAO figures for Basin countries' cultivated area (2008) significantly differ from the figures used by Deininger et al. 2011, especially for the Democratic Republic of Congo. FAO figures, in million hectares: Cameroon: 4.7; Central African Republic: 1.0; the Democratic Republic of Congo: 5.9; Gabon: 0.2; Equatorial Guinea: 0.1; the Republic of Congo: 0.3 (FAOSTAT. 2011. http://faostat.fao.org/, FAO, Rome; accessed December 2011).

---

the spared land would then sequester more carbon or emit less greenhouse gas (GHG) than do farmland. While this logic is attractive, models show that unless some accompanying measures are put in place, it may not systematically materialize.

The CongoBIOM model indeed indicates that land-production intensification, in a context of a growing demand for food as well as a huge potential for import substitution and an unlimited labor market (as in the Congo Basin), may lead to an expansion of the agricultural lands. Production costs fall, stimulating local consumption for agricultural products that rises above the level that can be reached only by the increase in productivity. The reduction in unit-production cost narrows the difference in opportunity costs that exists between agricultural and forest uses and generally more than compensates for the cost of converting forests into cropland. The productivity gains, by making the agricultural activities more profitable and attractive, thus can tend to amplify pressure on forested lands, which are generally the "easiest new lands" to access by farmers. Without pairing with accompanying policies and measures on land planning and monitoring, stimulating agricultural productivity will likely lead to more deforestation in the Basin (see chapter 3 on Recommendations).

International Demand Patterns: Indirect Impacts on the Congo Basin Forests

As the CongoBIOM model highlights, the Congo Basin could be affected by global trends in agricultural commodity trade despite its marginal contribution to global markets. The Basin is not yet fully integrated into global agricultural markets (with the exception of coffee and cocoa), but growing international demand for commodities could change this situation. The model illustrates how external shocks could indirectly affect the Congo Basin forests. Two policy shocks were tested under the CongoBIOM model: (S1) Increase in global meat demand by 15 percent by 2030; and (S2) Doubling of global demand for first-generation biofuels by 2030.

- Scenario 1 (S1): Increase in global meat demand. As living standards improve, diet patterns shift with an increase in animal calories consumption, particularly in emerging economies such as China, Russia, and India. If one considers that the average annual meat consumption in developed countries is 80 kilograms per capita and about 30 kilograms per capita in the developing world—and that the meat consumption in developing world is growing fast—then livestock production could increase sharply during the next decades. Such a trend creates a double pressure on climate change: the enteric fermentation of ruminants, which creates methane emissions, and pasture expansion or cropland expansion, to produce concentrated animal feeding, exert huge pressures on forests. During the last decade, Brazil has become a meat exporter; mechanized agriculture for soybean cultivation and intensive cattle grazing have been the dominant drivers of land clearing in the Amazon forest.[13]

  The Congo Basin has no comparative advantage for producing meat because it lacks the appropriate biophysical and climatic conditions for large-scale cattle farming. Yet, the increase in international demand for meat could impact the Basin's forest cover, as demonstrated by the CongoBIOM model (figure 2.7). Basin countries may suffer an indirect impact via the substitution of crops and the changing price signals. The model indicates that the development of cattle farming and feedstock production in Latin America and Asia may reduce crop production in these countries and that this reduction in supply may lead to a rise in crop prices. Congo Basin countries may

Figure 2.7 Channels of Transmission of Increase in Global Demand for Meat and Increase in Deforestation in the Congo Basin

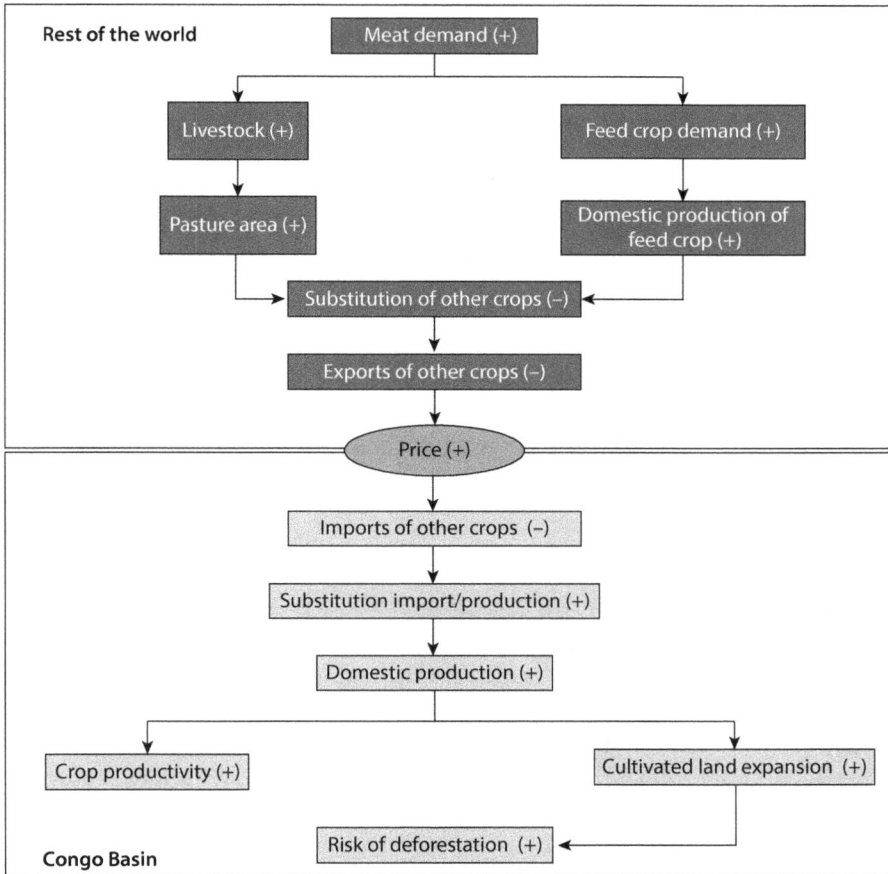

Source: Authors, adapted from IIASA 2011.
Note: + indicates increases and – indicates decreases.

react to this development by expanding the area under production for traditionally imported crops (corn in particular).

- Scenario 2 (S2): Doubling of first-generation biofuel production by 2030. Sugarcane and palm oil can be used directly to produce first-generation biofuels and are currently the predominant options in terms of biofuels.[14] There has been a spectacular increase in biofuel demand since 2000, primarily because of public sector support. This trend responds to the declining known and affordable reserves of fossil fuels and the need to diversify energy supply. While at some points, it was considered that an impact of substituting fossil fuels with biofuels could reduce the global carbon dioxide emissions in the atmosphere, this thinking is now being seriously questioned because of the potential contribution of biofuel development to increased deforestation in the tropics.

The climatic conditions are particularly suitable in tropical countries, and the planting of these first-generation biofuels directly competes with forest resources. However, as indicated above, despite the general trend of "land grabbing" elsewhere, the Congo Basin countries do not yet show significant signs of expanding new biofuel plantations, mainly because these countries lack a comparative advantage: other countries have large areas of suitable land with better performance in terms of infrastructure, productivity, and enabling business environments. The current trends in the Congo Basin are instead on the rehabilitation of the abandoned plantations.

That the Basin does not produce significant amounts of biofuels now does not mean that it will not eventually be impacted by the global expansion of biofuels. The modeling exercise conducted by CongoBIOM shows indeed that indirect impacts of biofuel expansions in other regions of the world will reduce agricultural exports from main exporting regions that could then increase deforestation in the Basin. These indirect impacts are comparable to those presented in figure 2.7.

Energy Sector[15]

Wood-Based Biomass Energy: The Largest Share of Energy Portfolio
In most African countries, the majority of energy is sourced from biomass. In Congo Basin countries (as in most African countries), the reliance on wood-based biomass energy from woodfuel and charcoal far exceeds that of other regions in the world. In 2006, it was estimated that in rural areas, 93 percent of the population in Sub-Saharan Africa depended on biomass resources for their primary cooking fuel and even in urban areas almost 60 percent of people used biomass for cooking (IEA 2006).

Woodfuel,[16] and particularly charcoal, demand continues to increase in Basin countries. Energy profiles vary from one country to another, based on country wealth, access to electricity, and availability and cost of wood and fossil fuel energy. Table 2.8 gives some key insights into the role of wood-based biomass energy in different countries in the Basin region. In the Democratic Republic of Congo, combustible renewables and waste (overwhelmingly woodfuel and charcoal) made up 93 percent of total energy use in 2008—in a context in which less than 12 percent of the population had access to electricity in 2009, and fossil fuels only accounted for 4 percent of the Democratic Republic of Congo's energy use in 2008. Kinshasa alone, with 8 to 10 million inhabitants, uses 5 million cubic meters of woodfuel, or equivalent, per year. In contrast, in Gabon the reliance on biomass energy is significantly lower, thanks to an extensive electricity network and subsidized gas for cooking.

Charcoal production in the Congo Basin region has more than doubled between 1990 and 2007, with an estimated 2.4 million metric tons of charcoal produced in 2007, almost 75 percent of it in the Democratic Republic of Congo. Cameroon follows, with nearly 10 percent of the region's charcoal production;

Table 2.8 Energy Consumption Portfolio and Access to Electricity in Congo Basin Countries, 2008 (and 2009)

Country	Energy use (kg of oil equivalent)	Energy use (kg of oil equivalent per cap.)	Combustible renewables and waste (% of total energy)	Fossil fuel energy consumption (% of total)	Electric power consumption (kWh per capita)	Access to electricity (% in 2009)
Cameroon	7,102	372.1	71	23.9	262.6	48.7
Central African Republic	—	—	—	—	—	—
Congo, Dem. Rep.	22,250	346.3	93.4	4	95.2	11.1
Congo, Rep.	1,368	378.4	51.3	43.5	150.2	37.1
Equatorial Guinea	—	—	—	—	—	—
Gabon	2,073	1,431.5	52.5	43.8	1,158	36.7

Source: World Bank, World Development Indicators database 2011; International Energy Agency (IEA) Electricity Access Database (http://en.openei.org/wiki/IEA-Electricity_Access_Database).
Note: Given that detailed data on wood-energy consumption is often not available, the numbers presented in this table deviate slightly from one document to another; however, general trends are usually confirmed by different data sources. kg = kilogram; kWh = kilowatt hour; — = not available.

however, because the country constitutes more than 20 percent of the region's population, per capita charcoal consumption is relatively low compared with other countries in the region (table 2.9).

Typically, the move to urban areas is associated with consumption switching from fuelwood to charcoal, the latter being cheaper and easier to transport and store.[17] Urbanization usually changes the way people consume energy; it is generally associated with a more energy-intensive lifestyle. Urban households are often smaller than those in rural areas, contributing to less efficient fuel use for cooking per capita. Besides being employed by households, charcoal is also often the primary cooking fuel in many small-scale restaurants and in kitchens of larger, public institutions, such as schools and universities, hospitals, and prisons. Because woodfuel is heavy and bulky—and thus difficult and costly to transport over longer distances—it is often converted into charcoal if it is to be used some distance from the forest where it was harvested.

Charcoal is mostly produced through traditional techniques, with low transformation efficiencies (figure 2.8). Earth-pit kilns or the slightly more efficient earth-mound kilns are used traditionally to produce charcoal in many parts of the world. In the former, wood is stacked in a pit, whereas in the latter it is stacked in a polygonal shape. In both cases, the wood is then covered with grass and sealed with a layer of soil before the kiln is lit. Both types of kiln yield only low-quality charcoal.

The biomass energy sector is a substantial contributor to the Basin countries' economies. The contribution of the wood-based biomass energy sector to the wider economy is estimated to be several hundred million dollars for most Sub-Saharan African countries. It is often considered the most vibrant informal sector, with the most value-added in Sub-Saharan Africa. The sector employs a significant workforce.[18] In most countries, transporters and/or wholesalers,

Table 2.9 Charcoal Production from Charcoal Plants

thousand metric tons

Country	1990	1995	2000	2005	2007
Cameroon	216	289	99	105	232
Central African Republic	0	0	21	120	182
Congo, Dem. Rep.	791	1,200	1,431	1,704	1,826
Congo, Rep.	77	102	137	165	181
Equatorial Guinea	—	—	—	—	—
Gabon	10	13	15	18	19
Total Congo Basin	1,094	1,604	1,704	2,112	2,440

Source: United Nations Statistics Division, The Energy Statistics Database (http://data.un.org/).
Note: — = not available.

Figure 2.8 Efficiencies of Alternative Kiln Technologies

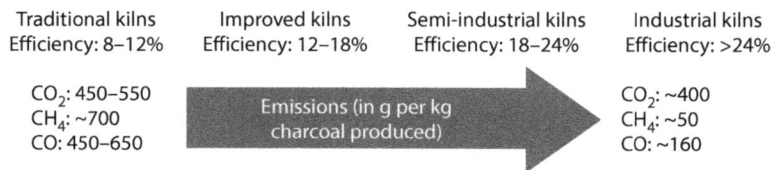

Traditional kilns
Efficiency: 8–12%

Improved kilns
Efficiency: 12–18%

Semi-industrial kilns
Efficiency: 18–24%

Industrial kilns
Efficiency: >24%

CO_2: 450–550
CH_4: ~700
CO: 450–650

Emissions (in g per kg charcoal produced)

CO_2: ~400
CH_4: ~50
CO: ~160

Source: Authors, based on Miranda et al. 2010.
Note: CO = carbon monoxide; CO_2 = carbon dioxide; CH_4 = methane; g = gram; kg = kilogram.

however, dominate the woodfuel supply chain and reap disproportionately large profits, leaving the producers with marginal benefits (the political economy of the charcoal trade network has been analyzed by Trefon et al. (2010) for Kinshasa and Lubumbashi, giving a detailed account of the different actors, including their strategies, relationships, and power, as illustrated in box 2.7). The sector's contribution to government revenues is limited due to widespread evasion of licensing fees and transport levies. National- and local-level governments are estimated to lose several tens or even hundreds of millions of dollars annually due to their failure to effectively govern the sector.

Regulations set up for the woodfuel sector tend be overly complicated, costly, bureaucratic, and often unenforceable, given the limited means available to local-level government representatives for executing their duties. The regulations burden mainly falls on producers, requiring them to engage in sustainable management of their forests. Most licensing bodies still operate as mere revenue-collection systems (a colonial-era heritage), without the number of licenses or quantities of harvestable wood licensed being linked to any kind of sustainability measures. Most of the time, these requirements are impossible to fulfill for various reasons, such as the incapacity to prove "land/tree ownership," costly preparation and implementation of sustainable forest management (SFM) plans, and bureaucratic processes to get any administrative from fiscal and/or forest administrations. The unrealistically complicated and costly process for

Box 2.7 Political Economy of the Charcoal Trade Network (Kinshasa and Lubumbashi)

The supply chain begins with charcoal producers who obtain (often temporary) access to trees through a process of negotiation involving tribal chiefs, private farmers, and, to a lesser extent, state officials. The specifics of gaining access to primary resources depend on whether the producers are local to the area and whether they are working with a group. Once the producers have obtained the right to cut some trees, charcoal kilns are fabricated. Since charcoal production provides a means of earning money with relatively few upfront investments (compared with agriculture), charcoal production has become an increasingly popular profession. During the dry season, farmers take up charcoal production to earn extra cash.

Charcoal producers have several selling options. In some cases, they take their charcoal to the city by bicycle, sometimes renting one for this purpose. If operating in a group, charcoal producers may designate one or several members to sell their product and share the profits. Roadside sales of charcoal also abound, with charcoal producers selling to passing drivers. These may be travelers returning to the city or professional traders who travel back and forth between the outskirts and city center to buy and sell charcoal (and other product). In Lubumbashi, it is common for these traders to operate by bicycle, whereas the greater distances in Kinshasa mean that trucks are generally employed. In rural areas, intermediaries collect charcoal produced in their region until they have enough to fill a truck, arranging a trader to pick up the supply. Trucks also frequently transport charcoal to Lubumbashi. Traders tend to form groups to share a truck and sell their charcoal to vendors in the large charcoal depots or to depot owners, who in turn sell it to smaller-scale retailers or those consumers who are able to buy entire bags of charcoal at once. The small-scale retailers then resell the charcoal on urban markets or street corners, where consumers purchase charcoal on a daily basis.

Source: Trefon et al. 2010.

producers to meet the regulatory requirements makes such informality the only solution. Typically, only small numbers of urban-based woodfuel traders are able to obtain exploitation permits, often resulting in an oligopolistic woodfuel industry. Strengthening existing laws and governance cannot provide a solution to the problems of the biomass energy supply chain. A profound reform of the policy and regulatory frameworks is necessary to "modernize" the sector (Miranda et al. 2010).

Woodfuel is largely underpriced. The pricing structure of woodfuel relies on incomplete consideration of the different costs all along the value chain and therefore sends contrary signals incompatible with sustainable forest-management practices. In most cases, the primary resource (wood) is taken as a "free" resource; uncontrolled open-access to forests tends to significantly undermine production costs. The larger part of the price is composed of transportation and retailing costs, which are down the supply chain. As a result, economic signals are insufficient to stimulate adoption of sustainable practices.

Limited Prospect for Significant Changes in Energy Profile

Wood-based biomass is likely to remain the major source of energy in future decades. In contrast to China and India, where the extent of wood-based biomass energy has peaked or will be peaking in the very near future, consumption of wood-based biomass energy is likely to remain at very high levels in Sub-Saharan Africa and may even continue to grow for the next few decades (see figure 2.9). Estimates in the *World Energy Outlook 2010* thus predict that by 2030, more than 900 million people in Sub-Saharan Africa may rely on wood-based biomass energy (IEA 2010).

Charcoal consumption in the Basin will remain at high levels or even grow in absolute terms over the next decades, based on prospects on population growth, increased urbanization, and relative price changes of alternative energy sources for cooking. High oil prices may prevent the poor from ascending the "energy ladder": in fact, it was anticipated that with rising income and stable prices, consumers would be able to move from woodfuel to charcoal and then to fossil fuels (that is, liquefied petroleum gas [LPG] or others). However, examples in different countries show that this phenomenon has not systematically materialized (Leach 1992). A regional study for southeast Africa estimated that charcoal consumption between 1990 and 2000 grew by about 80 percent in both Lusaka and Dar es Salaam (SEI 2002). Between 2001 and 2007, the number of households using charcoal for cooking in Dar es Salaam increased from 47 percent to

Figure 2.9 Number of People Relying on the Traditional Use of Biomass
millions

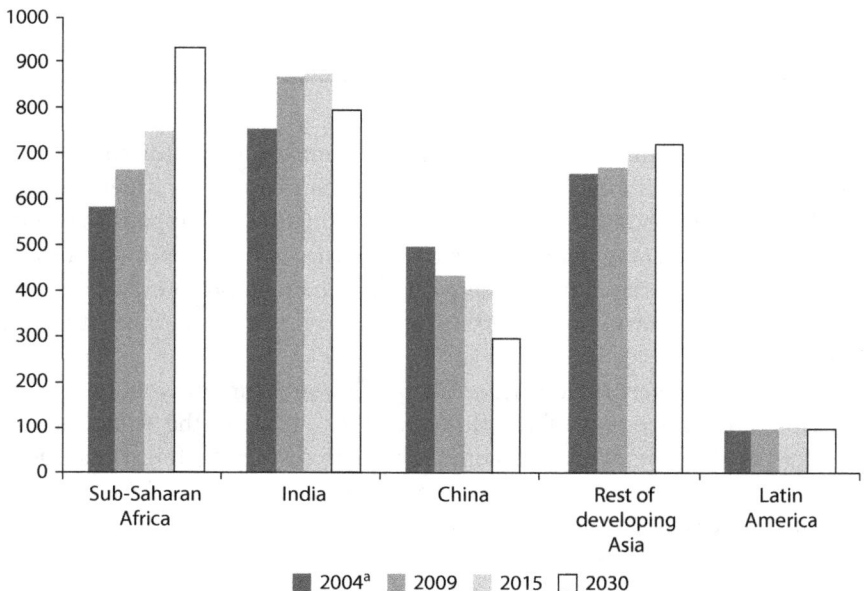

Source: International Energy Agency (IEA 2010; [a]IEA 2006).

71 percent, while the use of LPG declined from 43 percent to 12 percent (World Bank 2009). In Senegal, consumers made a massive return to using wood-based biomass for cooking after the elimination of subsidies caused prices for LPG to climb significantly. In some cases, rising fuel prices may even force wealthier segments of society to revert to wood-based fuels. In Madagascar, for example, the upper middle class—increasingly unable to afford LPG—has begun to switch back to charcoal. Supply reliability is another parameter that keeps consumers with wood-based biomass: not only can the purchased quantity be adjusted to the cash availability of the household, but wood-based biomass is also available through a wide network of retailers, and there is never a shortage of wood-based biomass. In contrast, the supply of other fuel options—especially LPG—has been reported by consumers to be unreliable and thus unattractive for regular use.

Impacts on Forests: Current and Future

Wood extraction for energy constitutes one of the major threats to forests in the Congo Basin, with a steady increase in wood removal in recent years. It is estimated that more than 90 percent of the total volume of wood harvested in the Basin is for woodfuel (see table 2.10) and that, on average, 1 cubic meter of woodfuel is required per person per year.

Table 2.10 Basic Data on the Wood Energy Sector in Central Africa

Food and Agriculture Organization of the United Nations (FAO) classification	Amount
Country	
Area (million ha)	529
Population (million inhabitants)	105
Forests	
Area (million ha)	236
Area (ha/inhabitant)	2.2
Standing stock	
Volume (m³/ha)	194
Total volume (million m³)	46,760
Biomass (m³/ha)	315
Total biomass (million m³)	74,199
Carbon (t/ha)	157
Total carbon (million t)	37,099
Production	
Wood energy (1,000 m³)	103,673
Industrial timber	12,979
Sawn wood	1,250
Some calculated ratios	
Consumption of wood energy (m³/inhabitant)	0.99
Production of wood energy/total woody production (%)	90

Source: Marien 2009.
Note: Estimates indicate that most of the removed wood is used as woodfuel; however, as the majority of woodfuel-collection activities are within the informal economy, the amount of wood removals may be underestimated.
ha = hectare; m³ = cubic meter; t = ton.

In rural areas, woodfuel consumption is no longer considered a leading direct cause of deforestation and forest degradation. Woodfuel used to be associated with energy poverty and forest depletion, a remnant of the "woodfuel crisis" era of the 1970s and 1980s (Hiemstra-van der Horst and Hovorka 2009). However, it is now acknowledged that demand for woodfuel in rural areas does not constitute a threat on natural resources. Analyses showed that a great portion of the woodfuel supply in rural areas comes from trees outside of forests, dead branches and logs, and even forest residues. When woodfuel is collected from natural forests, the regeneration capacity largely offsets the biomass withdrawals. It is thus seldom a primary source of forest degradation or forest loss.

Woodfuel collection becomes a serious threat to forests in densely populated areas. Woodfuel can be a severe cause of forest degradation and eventually deforestation when demanded by concentrated markets, such as urban household markets as well as industries and other businesses. In densely populated rural areas, the supply–consumption balance is usually defined by a high deficit for woodfuel, which creates huge pressure on forested areas surrounding the cities. With weak regulations and control over woodfuel harvesting, operators are likely to harvest as close to markets as they can in order to maximize profits, resulting in degradation in forest areas around urban markets (Angelsen et al. 2009). Similarly, given a potentially large demand within a small geographic area, industrial and other business demand for woodfuel can be a grave threat to local forest resources if not properly regulated.

Supply basins extend over time to satisfy the growing urban demand for energy, as illustrated in the case of Kinshasa. This megacity of 8 to 10 million inhabitants is located in a forest–savanna mosaic environment on the Batéké Plateau in the Democratic Republic of Congo. Kinshasa's wood energy supply of around 5 million cubic meters per year is mostly informally harvested from degraded forest galleries within a radius of 200 kilometers from the city. With gallery forests most affected by degradation from wood harvesting, even forests beyond the radius are experiencing gradual degradation, while the periurban area within a radius of 50 kilometers of Kinshasa has suffered total deforestation. Around the city, on the Batéké Plateau, plantations are also under implementation to diversify woodfuel supply (see box 2.8).

Charcoal, if not sourced properly, could represent the single biggest threat to the Congo Basin forests in the coming decades. It is unlikely that demand for charcoal will decrease; the main challenge thus relies on the countries' capacities to develop a sustainable supply chain for charcoal. This is particularly critical in densely populated areas. Plantations could supply wood-based biomass for energy. As an example, Pointe-Noire, located on the edge of a forest-savanna mosaic, is an industrial city with around 1 million inhabitants. The eucalyptus plantations (managed by the Eucalyptus Fibers Congo (EFC) firm), within 20 to 40 kilometers of the city, provide most of the woodfuel consumed in Pointe-Noire, while cost and transport considerations mean that

Box 2.8 Kinshasa: Toward a Diversification of Woodfuel Supply

Plantations developed around the megacity to help provide wood energy on a more sustainable basis. About 8,000 hectares of plantations were established in the late 1980s and early 1990s in Mampu, in the degraded savanna grasslands 140 kilometers from Kinshasa, to meet the city's charcoal needs. Today the plantation is managed in 25-hectare plots by 300 households in a crop rotation that takes advantage of acacia trees' nitrogen-fixing properties and residue from charcoal production to increase crop yields. Another scheme, run by a Congolese private company called Novacel, intercrops cassava with acacia trees in order to generate food, sustainable charcoal, and carbon credits. To date, about 1,500 hectares out of a projected 4,200 have been planted. The trees are not yet mature enough to produce charcoal, but cassava has been harvested, processed, and sold for several years. The company has also received some initial carbon payments. The project has been producing about 45 tons of cassava tubers per week and generates 30 full-time jobs plus 200 seasonal ones. Novacel reinvests part of its revenue from carbon credits into local social services, which include maintaining an elementary school and a health clinic.

the wood harvested from gallery forests up to 80 kilometers from Pointe-Noire tends to be converted into charcoal. Pointe-Noire's domestic energy supply is relatively sustainable, with little deforestation and degradation (Marien 2009).

There is an urgent need to modernize the sector. Corruption and oligopolistic marketing structures are major barriers to any attempt to formalize the woodfuel value chains. Existing regulations tend to limit access rights to local resources but are largely insufficient on the downstream levels of the value chain (transformation, transportation, and marketing). The political economy of this sector is very complex, with major vested interests. Most of the time, strengthening existing laws cannot provide solutions to the biomass-energy supply chain, and a profound reform of the policy and regulatory frameworks is necessary to "modernize" the sector (Miranda et al. 2010).

Transport Infrastructure Sector[19]

Transportation Networks: Insufficient and Deteriorated

The Congo Basin is one of the most poorly served areas in terms of transport infrastructure in the world, with dense tropical forests crisscrossed by several rivers, which, in turn, need numerous bridges. Given this complex environment, constructing transport infrastructure as well as properly maintaining it is certainly a major challenge for Basin countries. Recent studies indicate that investment required per kilometer of new roads is substantially higher here than in other regions of Sub-Saharan Africa, and the same scenario applies for maintenance.

Transport infrastructure assets are in poor condition. The physical capital of transport infrastructure has deteriorated in the Congo Basin, and this decline applies to the three main transport modes in the subregion: roads, waterways, and railways.

Road Transport Network: Sparse and Poorly Maintained

The road density in the six Basin countries is particularly low (between 17.3 kilometers per 1,000 square kilometers in the Democratic Republic of Congo and 71.7 kilometers per 1,000 square kilometers in Cameroon) compared to average Sub-Saharan Africa (149 kilometers per 1,000 square kilometers). However, the low road density in the Congo Basin seems partly offset by the low population density in most of its countries (and particularly in the rural areas; see figure 2.10 and figure 2.11).

Only a very limited ratio of the road network is considered of good quality. This ratio of classified roads in good and fair conditions range from 25 percent in the Republic of Congo to 68 percent in the Central African Republic,[20] which is globally lower than the average for the low-income countries (LICs) and the resource-rich countries (figure 2.12).

The road network has fallen into an advanced state of disrepair in most Congo Basin countries due to poor upkeep over the past decades (in some countries, the protracted civil war exacerbated this lack of maintenance). Until the past few years, the budget for road infrastructure had been excessively low, and the financial mechanisms (*Fonds routiers*) in place to support road maintenance

Figure 2.10 Total Road Network per Land Area
km/1,000 km²

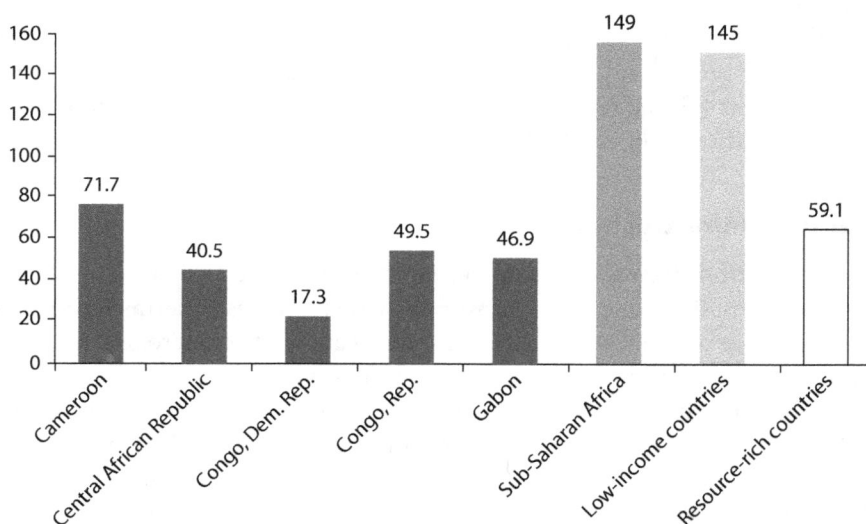

Source: Africa Infrastructure Country Diagnostic (AICD).
Note: km = kilometer, km² = square kilometer.

Figure 2.11 Total Road Network per Population

km/1,000 persons

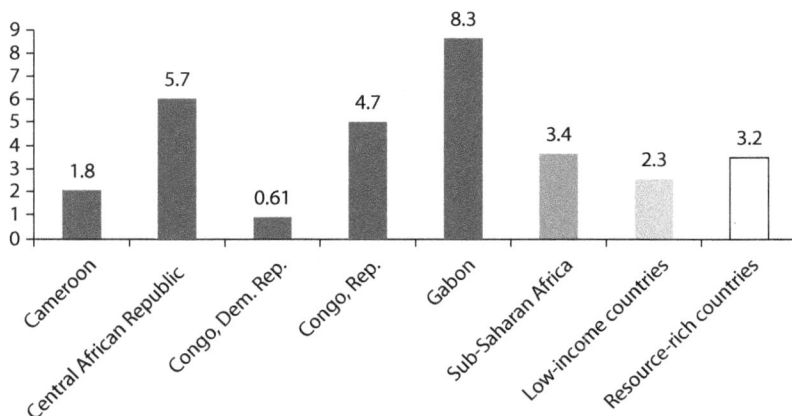

Source: Authors, prepared with data from Africa Infrastructure Country Diagnostic database (AICD) database (http://www.infrastructureafrica.org/tools; accessed in March 2012).
Note: km = kilometer.

Figure 2.12 Condition of Road Transport Infrastructure

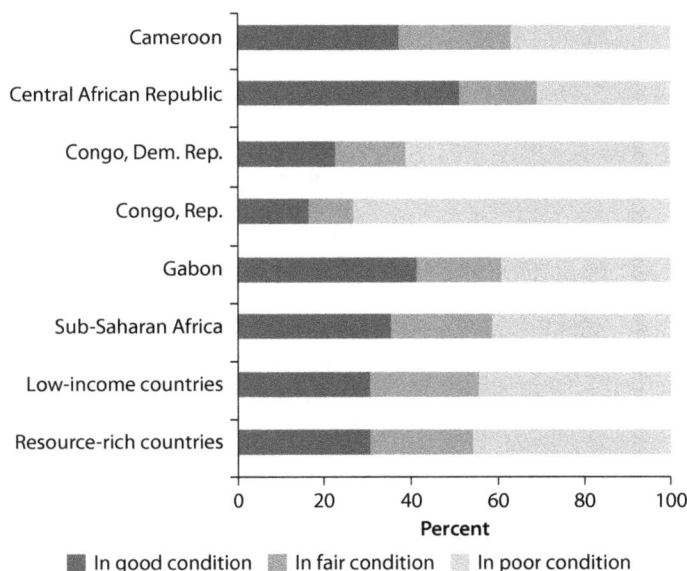

Source: Prepared with data from Africa Infrastructure Country Diagnostic (AICD) database (http://www.infrastructureafrica.org/tools; accessed in March 2012).

had been inadequate. As a result, the road transportation system in the Basin is characterized by poor quality of road transport, as illustrated by the Road Transport Quality index,[21] which was calculated for all Sub-Saharan African countries, normalized to 100 for the highest-quality road transport in South Africa (figure 2.13).

Deforestation Trends in the Congo Basin • http://dx.doi.org/10.1596/978-0-8213-9742-8

Figure 2.13 Road Transport Quality Index for Sub-Saharan African Countries

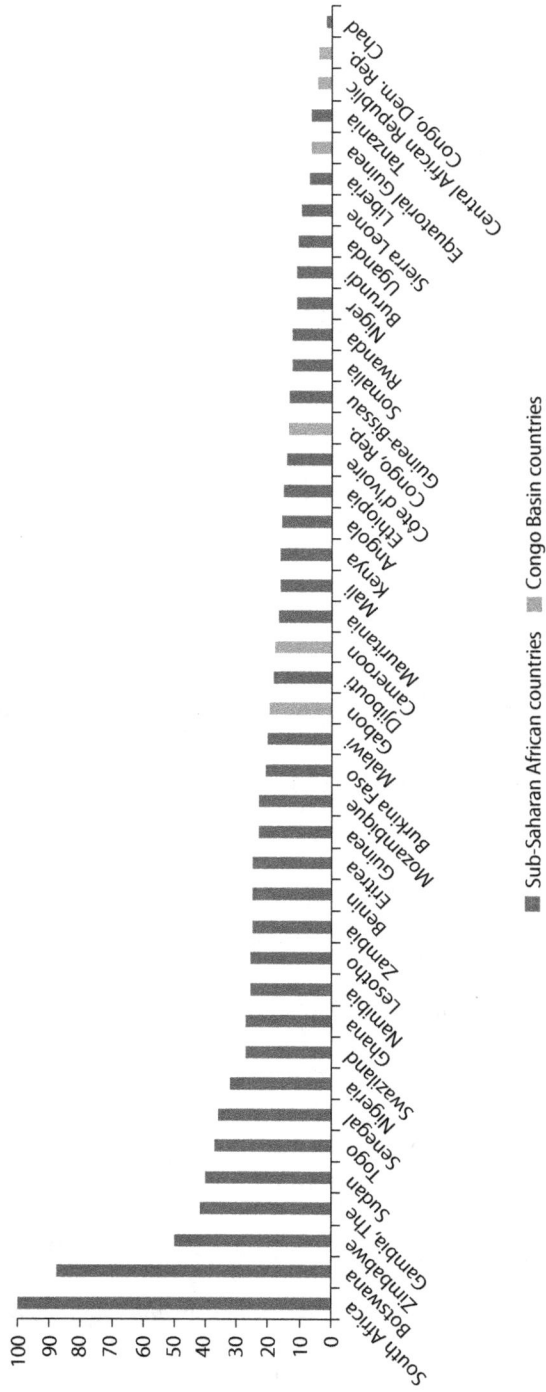

■ Sub-Saharan African countries   ■ Congo Basin countries

Source: Authors, prepared with data from Africa Infrastructure Country Diagnostic (AICD) database (http://www.infrastructureafrica.org/tools; accessed in March 2012).

River Transportation Network

Despite major potential, the waterway system remains a marginal transport mode in the Congo Basin, which has a navigable network of 12,000 kilometers covering nearly 4 million square kilometers in nine countries. The amount of goods—mainly agricultural products, wood, minerals, and fuel—transported via waterways is very modest, as this method is not usually reliable all year long. In principle, the system could significantly contribute to a multimodal transport network serving the region, particularly given low associated transport costs of US$0.05 per ton kilometer (vs. US$0.15 per ton kilometer for road or rail freight), albeit at significantly lower speeds. In practice, however, river transportation falls short of the contribution it could make to the overall economic development. River transportation has indeed declined since the 1950s; outdated and insufficient infrastructure, inadequate maintenance, poor regulatory frameworks, and numerous nonphysical barriers to movement, among other things, are to blame.

Railways Network

The railway network, essentially a legacy from the colonial era, has not been properly maintained and is largely disconnected. The total railway network in the Congo Basin countries is 7,580 kilometers, out of which more than a third are not fully operational. Railways in the Basin are mostly organized for the purpose of the extractive economy (timber and minerals; see table 2.11). Their development reflects the historically limited amount of intercountry trade in Africa. The railway network in the Basin is largely underperforming: it is comparatively better developed in the Democratic Republic of Congo and Cameroon.

Regional Corridors

Regional corridors have also deteriorated drastically. Only half of the major trade corridors in the Congo Basin countries are in good condition, leading to high costs for freight tariff, by far the highest cost in Sub-Saharan Africa (table 2.12). The quality of these corridor roads greatly impacts, in particular, the economy of the Central African Republic,[22] the only landlocked country in the Basin.

Table 2.11 Freight Composition as Percentage of Total Tonnage

Congo Basin country	Company	Timber	Cement and construction material	Fertilizers	Petroleum products	Ores and minerals	Agricultural products	Others	Total
Cameroon	Camrail	37	2	4	26	—	19	12	100
Congo, Dem. Rep.	CFMK	11	6	—	4	24	—	55	100
Congo, Dem. Rep.	SNCC	2	3	—	8	85	—	2	100
Congo, Rep.	CFCO	41	2	1	12	1	2	41	100
Gabon	SETRAG	30	—	—	—	60	—	10	100

Source: Africa Infrastructure Country Diagnostic (AICD) database, accessed in March 2012.
Note: — = not available.

Deforestation Trends in the Congo Basin • http://dx.doi.org/10.1596/978-0-8213-9742-8

Table 2.12 Africa's Key Transport Corridors for International Trade

Corridors	Length (km)	Roads in good conditions (%)	Trade density (US$ million/km)	Implicit speed	Freight tariff (US$/ton-km)
Western	2,050	72	8.2	6	0.08
Central	3,280	49	4.2	6.1	0.13
Eastern	2,845	82	5.7	8.1	0.07
Southern	5,000	100	27.9	11.6	0.05

Source: Africa Infrastructure Country Diagnostic (AICD) database, accessed in March 2012.
Note: km = kilometer.

Transportation Services: Expensive and Low Quality

Transportation prices in Africa are much higher than in anywhere else. Transport services in Central African countries are among the least performing in the world, with very high costs and low quality as measured by Logistic Performance Index (LPI) (figure 2.14). The Economic Community for Central African States (ECCAS)[23] reports that the freight transportation cost from Douala to Ndjaména is US$6,000 per ton and that the trip takes 60 days, while it is only US$1,000 and 30 days from Shanghai to Douala. The rail freight rate in the Democratic Republic of Congo is nearly three times the rate charged elsewhere in southern Africa.

Ambitious Plans to "Transform" Transport Infrastructure

The need to "transform Africa's infrastructure" is urgent in the Congo Basin. Ambitious plans are being prepared at the regional and continental levels, and individual countries are also now giving a much higher priority to prioritizing both infrastructure construction and rehabilitation.

Transport ranks high on governments' agenda. Over the past few years, most Basin countries have set transportation infrastructure goals to drive economic growth and development. In most cases, this push has translated into a major increase in their national budget allocations to the transport sector. A large portion is targeted to investments, as the top priority is really construction of new roads and, to a lesser extent, railways. Significant progress has also been made to mobilize external resources in order to support the reconstruction of the road network. In the Democratic Republic of Congo, for instance, the reconstruction of the road network was clearly marked as a top priority immediately after the armed conflict. Noteworthy efforts have been made to mobilize resources, and the Democratic Republic of Congo has secured major financial commitments from multilateral and bilateral donors as well as from China, as a result.

Programs are also created at regional and continental levels. Only a regional approach can address the highly fragmented network, and this approach could reduce the cost of infrastructure development and optimize the potentialities of the transport corridors within and outside the region. Accordingly, regional entities (ECCAS, Economic and Monetary Community of Central Africa [CEMAC], as well as the African Union) all aim to foster coordination among countries in Central Africa to unlock the potential of both extractive industries and

Figure 2.14 Logistic Performance Index (LPI) in Various Regions of the World

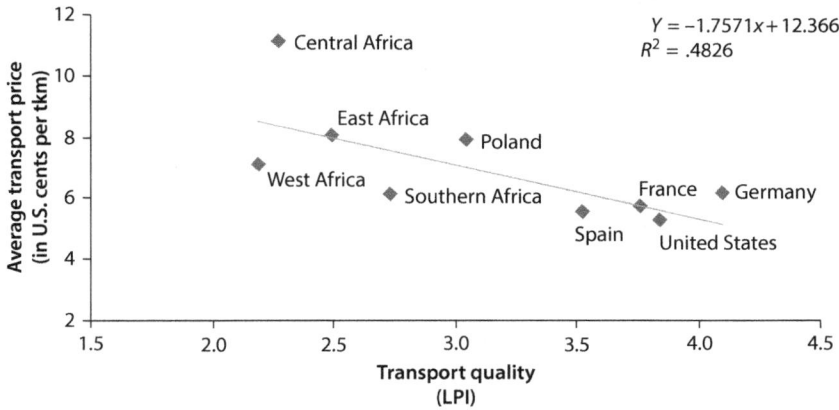

Source: Teravaninthorn and Raballand 2008.
Note: tkm = cost to move one ton one kilometer.

agricultural corridors. Examples of such initiatives include the Program for Infrastructure Development in Africa under the auspices of the African Union/ New Partnership for Africa's Development (NEPAD; at the continent level); the Consensual Road Network for Central Africa from the ECCAS; and others from other regional entities (CEMAC 2009).

Impacts on Forests: Current and Future

Transport infrastructure is one of the most robust predictors of tropical deforestation. Many studies have shown a positive correlation between road infrastructure development and deforestation in the various tropical forest blocks.[24] Roads accelerate forest fragmentation and reduce forest regrowth. The creation of transportation infrastructure (namely, roads and railways)[25] has both direct and indirect impacts on forests. Direct impacts are usually limited and only encompass a strip of a few meters on each side of the transport line (a security lane that needs to be deforested). Indirect impacts closely correlate with population density. In the case of the southern Cameroon, where population density is low, Mertens and Lambin (1997) showed that 80 percent of total deforestation occurs within a distance of less than 2 kilometers of roads; beyond a distance of 7.5 kilometers, deforestation ceases. This profile is different in more densely populated areas (as shown by many satellite images from the Congo Basin) and particularly in the transition zones. Overall long-term impacts in terms of deforestation could be of much greater magnitude and could extend over a long period of time, especially if forest governance is weak, local law enforcement is poor, and job opportunities in adjacent communities are limited.

The lack of infrastructure has so far "passively" protected Congo Basin forests. Most farmers are completely isolated from potential markets and are thereby cut off from participating in the broader economy that could foster competition and growth. This situation effectively creates landlocked economies even within countries with ports. This case is particularly evident in the Democratic Republic of Congo, with its considerable size and disconnected infrastructure. Farmers tend thus to rely on subsistence agriculture, with minimal impacts on the land. This situation greatly increases farmers' vulnerabilities to climatic shocks, but it also protects them from other external shocks (related to price volatility). Poor infrastructure has also been a major obstacle in the development of mining operations in the Congo Basin.

The CongoBIOM model has been used to assess the potential impacts of planned transportation developments for which funding has already been secured (box 2.9).

The CongoBIOM model presents the scenario "Transportation Infrastructure" as the most damaging on forest cover (see box 2.2). Among the multiple scenarios explored, scenario 3 (S3)—"Enhanced transportation infrastructure"—emerges as the one causing the highest rate of deforestation (see figure 2.3). The model shows that the total deforested area is three times

---

Box 2.9 Assumptions under the Scenario "Transport" under the CongoBIOM Model

Planned infrastructure: This scenario is based on the assumption that, with return of political stability and new economic potentialities (agriculture, mining, and so on), new transport infrastructure and the repair of existing infrastructure will increase by many fold. The model includes the projects for which funding is certain. This planned transport infrastructure information provided by the ministries for Cameroon, the Central African Republic, and Gabon, and by the World Bank for the Democratic Republic of Congo and Republic of Congo (AICD), was used in the model to forecast the impact.

Population density: As illustrated in the section on "Impacts of Transport on Forests" in this chapter, the preservation of the forest cover along transport axes largely depends on population density. It is critical to include the distribution of the population as well as the prospects for growth during the simulation period, because with the increase in population, dynamics of forest access and resource extraction will change. Therefore, population growth parameters were integrated in the CongoBIOM model. In the Basin, population is expected to double between 2000 and 2030, with an average annual growth rate of 3.6 percent between 2000 and 2010, and 2.2 percent between 2020 and 2030, leading to a total population of 170 million by 2030.

Urbanization trends have also been computerized: As in other developing regions, the urbanization process is expected to intensify in the Congo Basin. From United Nations estimates (2009), the number of cities in the Congo Basin with more than 1 million inhabitants should jump from four in 2000 to eight in 2025, with 15 million inhabitants in the city of Kinshasa alone. North and southwest Cameroon and the eastern Democratic Republic of Congo border will continue to register high population densities.

Source: International Institute for Applied Systems Analysis [IIASA] 2011.

higher than in the business-as-usual scenario (and total GHG emissions more than four times higher, as most of the deforestation, according to the model, occurs in dense forest). Most of the impacts do not result from the infrastructure development itself but from the indirect association with the fact that economic potentials are unlocked through an enhanced access to markets and higher connectivity.

Enhanced transportation lowers transportation costs. The CongoBIOM model computerizes current transportation costs throughout the Basin into a spatially explicit data set. The internal transportation cost has been estimated on the basis of the average time needed to go from each simulation unit to the closest city with more than 300,000 people in 2000 (including cities in neighboring countries), based on the existing transportation network including roads, railways, and navigable rivers; elevation; slope; boundaries; and land cover. S3 uses the same methodology and same parameters and computerizes the internal costs with planned infrastructure for the 2020–30 simulation period. The transportation costs are expected to reduce in the same magnitude as the transportation time.[26]

Improved infrastructure unlocks agriculture potential. Reduced transportation costs can lead to substantial changes in the economic equilibrium of a rural area and the dynamic of agricultural development. The causal chain that the model highlighted is as follows:

Improved Infrastructure → Increase in Agriculture Production
→ Increased Pressure on Forests

Enhancing transportation networks tends to reduce the price of the agricultural product to the consumer while producer prices net of transport costs tend to increase. This leads to a rise in consumption (often through substitution phenomenon),[27] which, in turn, encourages producers to yield more. Typically, a new equilibrium would be reached, with a larger volume and lower price compared to the initial situation. Under the scenario "Improved infrastructure,"[28] the CongoBIOM model projects a 12 percent increase in the total volume of crops produced and a decrease in the price index for local crops, following infrastructure improvements in the Basin. The graphs below show the projected deforestation "hot spots" due to agricultural expansion.

The international competitiveness of agricultural and forestry products could also benefit from reducing transportation costs. However, despite huge potential in terms of land availability and suitability for biofuels, the CongoBIOM model simulation indicates that the poor business climate would still place Basin countries in a disadvantaged position in comparison to other large basins. Nonetheless, signals of growing interest from foreign investors in Cameroon seem to suggest that new trends could be foreseen, potentially related to the policy decisions taken by other large-producer countries (for example, a moratorium in Indonesia on any new oil palm plantation leading to deforestation; see box 2.10).

Box 2.10 Palm Oil in Cameroon: A New Impetus?

Worldwide demand for palm oil, the number 1 vegetable oil globally, is projected to rise as the world looks for affordable sources of food and energy. In 2011, Malaysia and Indonesia dominated the production of palm oil, but strong consumption trends have made it an attractive sector for investors seeking to diversify supply sources across the tropics, including the Congo Basin. A case in point is Cameroon, where at least six companies are reported to be trying to secure more than 1 million hectares of land for the production of palm oil (Hoyle and Levang 2012). In 2010, Cameroon produced 230,000 tons of crude palm oil across an estate of 190,000 hectares (independent smallholdings accounted for 100,000 hectares; supervised smallholder plantations and agroindustrial plantations accounted for the balance) and was the world's 13th-largest producer. Compared to other crops in the Congo Basin, where productivity tends to trail far behind other countries' performance, palm oil yields in Cameroon are also among the highest in the world (on a par with Malaysia's). Because of its potential in terms of growth, employment, and poverty reduction, industrial palm oil production is a national priority, with plans to increase production to 450,000 tons by 2020. Some of the sites preidentified in emerging land deals could be problematic because the proposed plantation sites appear to be in high conservation value forests or near biodiversity hot spots.

In many areas, the opening of new roads is immediately associated with an increase in unlawful activities and, particularly, illegal logging. The domestic demand for wood (both for construction and energy), while long overlooked, is now recognized as more important than the supply to international markets; this leads to increased pressure on forest resources and—without a proper governance system—uncontrolled activities tend to soar.

Logging Sector

A Formal/Informal Dual Profile

The Congo Basin's logging sector has a dualistic configuration with a highly visible formal sector that is quasi-exclusively export oriented and dominated by large industrial groups with foreign capital; its informal sector has long been overlooked and underestimated. Domestic timber production and trade are usually unrecorded, and thus little information is available on the scope of the domestic sector.

Formal Sector: Major Progress, to be Pursued

The formal sector has made huge progress over the last decades. The industrial logging sector has been a major economic sector in most Congo Basin countries. It represents the most extensive land use in Central Africa, with about 45,000 square kilometers of forest under concession. Basin countries have made major progress on SFM in logging concessions in Central Africa over the past decades. The region is one of the most advanced in terms of areas with an approved (or under preparation) management plan (MP).

However, progress needs to be pursued to fully operationalize SFM principles on the ground. Studies indicate that, despite this progress, SFM principles are yet to fully materialize on the ground in the logging concessions. While a lot of technical expertise is usually put in the preparation/approval process of the MP for a logging concession, it seems that in many countries much less attention is given to implementation of the plan. In addition, current standards on SFM were based on the knowledge of forest dynamics at the time of the elaboration of the regulations and would benefit from a revision. In fact, practical knowledge has been accumulated at the level of the concessions over the past decade, and there is an opportunity to adjust the SFM parameters and criteria. This revision could also take into account new elements, such as the climate change that is already affecting forest dynamics in the Basin (growth/mortality/regeneration rates). Moreover, the SFM principles could move away from a purely technical approach and more broadly include nontimber products, biodiversity conservation, and environmental services as part of forest MPs. Such multiuse forest management would better respond to the needs of the multiple stakeholders dependent on forest resources and also add value to the forest. SFM could serve as a tool toward a multiuse management approach, while planning for multiple uses would be elevated to the landscape level.

Informal Sector: Long Overlooked

The informal timber sector has long been overlooked both by the national entities and by the international community, which, over the past decades, mostly focused their attention on the industrial and export-oriented sectors. In 1994, the devaluation of the regional currency (African Financial Community Franc, FCFA) boosted timber exportation at the expense of the domestic market, which, as a consequence, largely contracted. The recovery and boom of the domestic market in recent years is a sharp turnaround, and the domestic and regional timber economy is now recognized as just as important as the formalized sector.

In some countries, the economic importance of the informal sector is assumed to exceed the formal sector (Lescuyer et al. 2012). For example, in Cameroon and the Democratic Republic of Congo, domestic timber production already surpasses formal timber production; in the Republic of Congo, domestic timber production represents more than 30 percent of total timber production. Only recently, research work on the informal sector substantiated its importance in terms of both estimated timber volumes and the number of jobs associated with informal activities (from production to marketing). Domestic operators are now recognized to serve as engines for small- and medium-enterprise development (Cerutti and Lescuyer 2011; Lescuyer et al. 2012).

The informal sector is powered by booming domestic and regional markets. Demand for timber has been soaring on local markets to meet the growing needs of urban populations. However, while focus has been put, so far, on export trends (to European markets as well as to Asian markets), very few information exists on the rapidly growing domestic markets in the subregion (both national and regional). Research shows that demand for construction timber rapidly increases from urbanizing cities. This demand comes from growing urban centers in the Basin but also expands way beyond. It was recently documented that well-established transnational timber supply networks from Central Africa to as far as Niger, Chad, Sudan, the Arab Republic of Egypt, Libya, and Algeria have driven the growing urban demand for construction material (Langbour, Roda, and Koff 2010). However, knowledge and understanding of these markets (such as type of products, volumes, prices, and flows) remain partial and limited.

The informal sector provides considerable socioeconomic benefits to local communities. The informal sector offers financial contributions to rural economies that are largely ignored in official statistics; recent research shows that the informal sector provides for much higher direct and indirect local employment than does the formal sector, with benefits more equally redistributed at the local level than have been achieved through formal sector activities. Lescuyer et al. (2010a) estimated the financial gain generated by the informal sector (that is, based on aggregated local wages, fees, and profits) at around US$60 million per year for Cameroon, $12.8 million for the Republic of Congo, $5.4 million for Gabon (Libreville area only), and $1.3 million for the Central African Republic (Bangui area only). Generally, the socioeconomic benefits generated by chain saw milling are distributed more widely in communities than are benefits created by conventional logging. The same study also demonstrated that the revenue from chain saw milling that remains in rural economies in Cameroon is four times as high as the area fee, the latter being the tax paid by industrial logging companies and redistributed to local councils and communities. Further, the income gained from chain saw milling activities also stimulates a secondary economy, thus providing further benefits as secondary service and trade activities develop.

Despite these important local socioeconomic benefits, current regulatory frameworks fail to properly regulate domestic timber production. Because of the quasi-exclusive focus on industrial timber sector, forest-related laws and regulations prepared since the 1990s have been designed with a clear bias on industrial operations, with little attention to smaller operations. Consequently, legal/ regulatory frameworks are not adapted to small forest enterprises, which are thus constrained to illegality with a greater adverse impact on natural forest resources (due to overexploitation of timber resources by informal operators). As long as national and international policy makers continue to largely disregard local timber production and consumption—and as long as there is no clear framework that regulates domestic timber production and trade—there is little prospect that illegal timber trade can be reduced. There is an urgent need to focus the efforts on the formalization of the informal sector and define new rules and regulations that can support the sustainable development of this vibrant sector while preserving the capital of natural forests.

Left unregulated, the informal sector has been "captured" by vested interests and its socioeconomic benefits have been compromised by corrupt practices. The informal sector involves a large number of operators including sawyers, porters, retailers, firewood traders, mill owners, log transporters, and others. Despite operating outside of governance and legal schemes, it significantly interacts with national entities (forestry administration, customs, finance, and so on). A large share of the benefits is captured by community elites, individuals in the lower level of the supply chain (that is, traders), or corrupt government officials seeking informal fees. These "unofficial" payments to government officials and local elites could also be considered lost revenue to the state. Lescuyer et al. (2010b) extrapolated estimates of these payments to the overall volume of approximate informal production and thus calculated revenue losses from the informal sector to US$8.6 million in Cameroon, $2.2 million in the Republic of Congo, $0.6 million in the Central African Republic, and $0.1 million in Gabon.

The Challenge to Meet a Growing Demand for Timber (International and Domestic)

International Demand

Congo Basin countries are a relatively small player in terms of timber production at the global level. With an average production of 8 million cubic meters per year, Central African countries produce about 80 percent of the total volume of African timber; however, its contribution to international timber production remains low. Central Africa trails far behind the other two major tropical forest regions in terms of tropical timber production, with only 3 percent of global production of tropical round wood and just 0.4 percent global production of round wood.[29]

Contribution of Congo Basin countries to processed timber production is even lower. A global analysis of trade in further-processed timber shows that the value of exports for all International Tropical Timber Organization (ITTO) producer countries combined was about US$5 billion in 2000, 83 percent of which originated in countries in Southeast Asia, 16 percent in Latin America, and only 1 percent in Africa. Of African countries, Ghana and Côte d'Ivoire alone contribute almost 80 percent to the further-processed timber trade, meaning that the share from Central Africa is very small (Blaser et al. 2011).

Asian markets absorb more and more timber exports from the Congo Basin. Europe used to be the traditional market for Basin timber countries. While still important, it tends to contract while Asian markets expand. In the late 2000s, when timber demand from the European Union almost collapsed with the economic crisis, China's demand proved to be more resilient and helped to sustain Central African timber exports during recent years (box 2.11). Asian markets also present different profiles and preferences, which could eventually change the way timber is produced in Central Africa. Asia is now the main exportation hub, receiving about 60 percent of total exports during the period 2005 to 2008. It strengthened its position in 2009, at the height of the crisis, by exceeding 70 percent of total exports. In addition Asia—and particularly China—imports a broader selection and higher volume of lesser-known secondary species, which

Box 2.11 On Timber Trade with China and Other Emerging Asian Markets

To better grasp the influence of China and other emerging Asian markets on timber management and exports in the Congo Basin, it is important to understand the market dynamics specific to the timber trade between the Central African countries and China. Following the 1997 Asian crisis, the timber demand of Asian countries, most notably China, grew rapidly. Between 1997 and 2006, China's total timber product imports almost quadrupled in volume (roundwood equivalent) from approximately 12.5 million cubic meters to more than 45 million cubic meters. China is now the leading importer of timber products in the world.

With China's manufacturing sectors rapidly expanding, the demand for unprocessed timber is skyrocketing. This is also reflected in the changing composition of timber imports by China. Through the 1990s, China mainly imported large quantities of plywood, but the substantial increase in timber imports over the last decade is almost exclusively based on increased log imports, while sawn wood imports stagnated and plywood imports actually decreased. Accordingly, China as has been the top destination of logs exported from the Basin for several years now, surpassing historical destinations such as Italy, Spain, or France (Blaser et al. 2011). For more than 10 years, Gabon has been the largest Central African supplier of logs to China (for example, with exports worth US$400 million in 2008), followed by the Republic of Congo, Equatorial Guinea, and Cameroon.[a] In comparison to exports from other Congo Basin countries, the Democratic Republic of Congo's official timber exports to China remain at less than $20 million. However, the Democratic Republic of Congo's timber sales to China have been trending sharply upward and the volume of timber illegally shipped through bordering countries has not been quantified, making the Democratic Republic of Congo's timber export sector worth closer investigation.

In line with the above export trends, several Western logging companies that have been operating in Africa for decades have recently been taken over by investors from China and other emerging Asian countries. For example, the formerly French, then Portuguese, firm Leroy-Gabon, was taken over by Chinese interests. The originally French, then German (from 1968), then Danish (from 2006), company CIB operating in the Republic of Congo was sold to the Singapore-based firm Olam International (controlled by Indian investors) by the end of 2010.

a. The effect of Gabon's 2010 log export ban on China's imports of Gabonese timber has yet to be fully examined.

may become more important as the stock of primary export species degrades or gets more costly to access in remote forest areas.

Domestic (and Regional) Demand

Domestic demand for construction timber is booming and is currently quasi-exclusively supplied by an unregulated and underperforming sector. This trend is unlikely to fade, as most Congo Basin countries experience a strong urbanization process. In addition, as indicated above, demand for informal timber comes from other African countries (such as Niger, Chad, Sudan, Egypt, Libya, and Algeria) where demographic growth and urbanization are considerable.

Sources for timber need to be regulated and diversified. The current situation already generates major inefficiency in the provision of timber to domestic markets as well as huge pressures on natural forests. Unless the domestic timber supply becomes properly regulated, this situation will exacerbate and cause major adverse environmental impacts. The concept of "community forestry" has been embraced by most Basin countries and is now reflected in their legal framework; however, challenges remain in terms of operationalization of this concept. Plantations and agroforestry systems could also contribute to the diversification of the timber supply for domestic markets. In addition to better management of natural forests, the need for alternative timber sources seems clear.

Processing Capacities: Still to be Modernized

One of the paradoxes of Central Africa is that net trade in timber furniture is negative, with imports totaling US$16.5 million against US$9.5 million in exports. At first glance, the fact that countries like Cameroon are net importers of furniture may seem paradoxical. However, because the import volume and demand for high-end furniture is mainly driven by urban elites, hotels, restaurants, and administration, local producers find it difficult to tap into this sizable market due to quality and design constraints as well as a lack of appropriate equipment and skills. Therefore, the lower quality of locally manufactured furniture prevents local manufacturers from competing with global furniture manufacturing to meet the domestic demand for higher-end furniture.

Processing capacities in the Congo Basin, when they exist, are essentially limited to primary processing: sawn wood, and peeling and slicing for the production of plywood and veneer (that is, primary processing). Accordingly, more than 80 percent of timber-processing facilities in Central Africa are saw-mills[30] (figure 2.15). Together, Cameroon and Gabon represent 60 percent of the subregional processing capacity. In most Central African countries, secondary or tertiary timber processing—that is, the stages that generate the most added value and employment, such as the manufacture of molding, flooring, furniture, and joinery—is in an embryonic stage, while the industry is more developed in West Africa (Ghana, Côte d'Ivoire, and Nigeria). Overall, Basin countries lag behind except in the production of moldings, floorings, and other dry and profiled timber, which has expanded in Cameroon during the past decade.

Techniques used by the informal operators are largely inefficient (hands-free chain sawmilling), but the low-priced domestic market conditions tend to counter any attempt to improve processing methods. Various studies show that small-scale forest operations are largely inefficient, with a low processing rate, and that they tend to gain in profitability with a certain scale. Furthermore, the informal sector supplies markets that are less selective than export markets; chain saw operators tend to be much less efficient in their use of trees to produce timber; and informal activities usually over-log the most accessible areas, surpassing regeneration rates.

Having a higher performing and modern timber processing industry has always been a top priority for the Congo Basin governments. Progress on that front has been minimal so far, but signs suggest that this is likely to change in the

Figure 2.15 Timber-Processing Units in Central Africa, 1975 and 1995

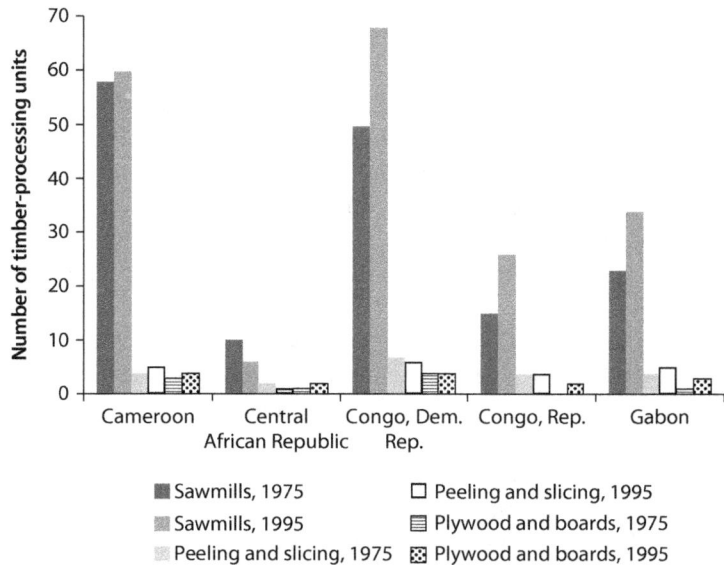

Source: Authors, prepared with data from International Tropical Timber Organization (ITTO) 2006.

coming years; however, governments are putting more pressure on the operators to maximize the processing level and to increase in-country value addition and employment. Log export restrictions are currently being applied in the form of (partial) export bans or the setting of local log-processing quota (minimum processing quota). Ambitious steps are being taken by Gabon's government, for example, to develop a free-zone area—special economic zone, or SEZ, in Nkok, about 30 kilometers from Libreville—on 1,125 hectares in partnership with the private operator Olam.[31]

Further developing the processing subsector for the domestic market could expand marketing options to lesser-known secondary species. Although the value of sawn wood, veneer, plywood, and flooring products depends on the species used, manufactured timber products are not necessarily species specific and instead their sales value rests more on their appearance and quality. Secondary species represent a growing percentage of authorized logging due to the great diversity of species in Basin forests and the degradation of residual primary forests. Developing the industry for manufactured wood products could thus add value to secondary species and further support larger-scale acceptance of these lesser-known species in future timber supplies. Many secondary species of Central Africa lend themselves well for further processing and are of interest due to the the their excellent technical qualities as well as their broad availability.

Impacts on Forests: Current and Future
Unlike in other tropical regions, in the Congo Basin logging activities usually do not entail a transition to another land use. Logging activities in the Basin generally

lead to forest degradation rather than deforestation.[32] Limited cumulative figures are available on the specific degradation impacts of logging activities. Annual degradation rates for dense forest in the Basin have been estimated at 0.09 percent based on gross degradation of 0.15 percent combined with recovery of 0.06 percent (Nasi et al. 2009). Although available data on the quantitative impact of logging operations on biomass and carbon stocks are insufficient, the GHG emissions from industrial logging activities are considered to be low. The impact may, however, be more significant for informal logging activities that do not apply minimum standards to manage the resources and thus tend to be more damaging to the forests.

Impacts from Industrial Logging Sector

As indicated in chapter 1, the industrial logging sector in the Congo Basin has two specific features that tend to drastically limit its impact on forest carbon[33]: adoption of SFMs (and, in some cases, subscription to forest certification schemes) and the high selectivity in valued species. In a convention industrial concession, it is estimated that for the first round of logging in old-growth forests, the total disturbed area accounts for approximately 5.5 percent of total forest area. Second or third rotation logging in logged-over forests increases the damage to over 6.5 percent of the total surface. Box 2.12 describes the impacts of the different logging operations.

Field studies on the carbon impact of selective logging in the Republic of Congo quantify the impact of selective logging on forest carbon stock (Brown et al. 2005). The research estimates that the carbon impact from the test logging site totals 10.2 tons of C per hectare of concession area, or a total carbon impact of 12,174 tons of C for a total of 3,542 tons of C of extracted biomass carbon (that is, the commercial log). This figure represents a comparatively low total carbon impact, which breaks down roughly as follows: 29 percent extracted biomass carbon, 45 percent damaged biomass carbon in logging gap, 1 percent damaged biomass carbon in skid trails, and 25 percent biomass carbon impact of logging roads.

Research also suggests that such selective logging activities have comparatively less impact on carbon stock than do reduced-impact logging activities in the Amazon Basin. There are several explanations for the low impact of the case study of highly selective logging in the Congo Basin. In the Amazon test sites, the ubiquitous presence of lianas resulted in more damage to the areas surrounding the extracted trees; there was no presence of lianas at the test site in the Republic of Congo. Further, the overall biomass and thus carbon impact of highly selective logging in the Congo Basin is low due to the relative proportions of the extracted timber trees.

In addition to the limited impact during logging activities, industrial concessions are generally managed under rotation cycles, meaning that the plots will not be logged again before a 20- to 30-year period, leaving enough time for biomass to regenerate. As a result, it is expected that under proper implementation of SFM principles, a concession should globally maintain a carbon stock over the long run (see box 1.1 in chapter 1).

Box 2.12 Typical Impact of Commercial Logging Operations

- Logging base camp: 0.03 to 0.1 percent of forest cover of the concession area is cleared for the purpose of the base camp/camps, according to companies. However, subsequent to the establishment of a base camp, the pressure on the surrounding forest increases rapidly due to agricultural activities, hunting, and so on. Little quantitative data are available on the extent of indirect impact from logging base camps.

- Logging access roads: Creation of logging roads involves the clearing of a strip of forest and the compacting of the soil. Access roads are typically 4 to 25 meters wide. Primary and secondary roads generally account for 1–2 percent of surface disturbance (including the road edges that are also cleared).

- Incidental damage: The felling of trees also contributes to damage and uprooting of adjacent trees and vegetation in the logging plot, including total damage of trees as well as broken-off branches of nearby trees. As part of an operation with an extraction intensity of 0.5 trees per hectare, one generally estimates that per one square meter of extracted timber, damage is caused to 4.3 square meters of surrounding forest area. Vine cutting prior to felling significantly reduces the impact.

- Skidding trails: Skidding trails create the least impact of the different factors, in particular in Africa, where extraction is highly selective. The track that is opened is usually rapidly overgrown, large trees are avoided during development of the track, and skid trails are often undetectable from aerial photographs shortly after operation. As part of an operation with an extraction intensity of 0.5 to 1 tree per hectare (5–15 m^3/ha), one generally estimates that about 3 percent of the forest floor is covered by skid trails—half of the area caused by the actual extraction.

- Log dock: The log dock is an opening in the forest to accommodate temporary storage of extracted logs prior to further road transport. This usually accounts for 0.3 percent of the total surface used.

Impacts from Informal Logging Sector

The major threat from logging activities is expected to come from the informal sector that supplies the boiling domestic market. Although the ecological impacts and sustainability of the informal timber sector have not been scientifically established, experts suggest that the informal chain saw milling industry tends to lead to depletion of forest resources due to the combination of several factors:

- The informal-sector supply markets are less selective than the export markets, as such the extraction rate from logging activities is considered to be higher: reducing selectivity and increasing the number of secondary species in the market generally boost the ecological impact per logged area.

- The processing rate of the chain saw industry is very low, requiring many more resources for the same volume of processed products.
- The informal activities are not governed by logging cycles and tend to over-log the most accessible areas (closed to markets or transportation access); this leads to a progressive erosion of the resources, as the regeneration rate cannot cope with extraction rates.

As long as the informal sector is left unregulated, its impacts on natural forests are expected to rise and progressively degrade forests in most highly populated areas.

Mining Sector

Mining Resources: Abundant but Still Largely Untapped

The Congo Basin is home to a vast wealth of various valuable mineral resources. Mineral resources span from metals (copper, cobalt, tin, uranium, iron, titanium, coltan, niobium, and manganese) to nonmetals (precious stones, phosphates, and coal) and other mineral resources (table 2.13). Mining activities (industrial and/or artisanal) are present at several locations in the Basin countries, but the following major mineral provinces stand out: the cupriferous belt of Katanga; the auriferous province located in the Democratic Republic of Congo; the bauxite (aluminum) province located in the central-north region of Cameroon; the iron province located at the borderline between Cameroon, Gabon, and the Republic of Congo; and the nickel and cobalt province of Cameroon.

With the exception of the Democratic Republic of Congo, which has both the richest deposits and a long history of mining (mainly in the southeast), the vast mineral wealth of the Congo Basin remains largely underdeveloped.

Table 2.13 Common Minerals in Congo Basin Countries

Mineral	Country
Gold	Cameroon, Central African Republic, Democratic Republic of Congo, Republic of Congo, Equatorial Guinea, Gabon
Diamond	Cameroon, Central African Republic, Democratic Republic of Congo, Republic of Congo, Gabon
Iron	Cameroon, Democratic Republic of Congo, Republic of Congo, Gabon
Uranium	Democratic Republic of Congo, Republic of Congo, Gabon
Lead	Democratic Republic of Congo, Republic of Congo, Gabon
Tin	Cameroon, Democratic Republic of Congo, Republic of Congo
Aluminum	Cameroon, Democratic Republic of Congo, Republic of Congo
Manganese	Democratic Republic of Congo, Gabon
Copper	Democratic Republic of Congo, Republic of Congo
Titanium	Cameroon, Republic of Congo
Cobalt	Cameroon, Democratic Republic of Congo
Niobium	Gabon

Source: Reed and Miranda 2007.

Many factors have hampered the development of the mining sector:

- Civil unrest over the past two decades: The region has experienced numerous rebellions and civil conflicts, resulting in a highly unstable and risky environment for business. Amid such a climate of instability, any investment capital flowed out of the region. The Republic of Congo, the Central African Republic, and the Democratic Republic of Congo have also struggled with armed groups who have often used mineral wealth as a source of funding for their activities.

- Lack of infrastructure: Infrastructure assets (including transportation) are highly insufficient and deteriorated in the Basin countries. This situation results from a lack of investments in the infrastructure but is also a direct consequence of civil unrest and armed conflicts. For example, in the Democratic Republic of Congo, sporadic looting and two periods of armed conflict have destroyed much of the infrastructure. The lack of reliable transportation infrastructure has until now been a major setback to the exploitation of minerals resources in the Congo Basin countries.

- Due to poor governance, the climate of investment is also not conducive to doing business. Additionally, complex and often arbitrary, and predatory, taxation discourages investment (World Bank 2010).

- A heavy reliance on oil: Oil booms and subsequent "Dutch disease" syndrome have distracted the government in most of the Basin countries from economic diversification. For instance, in Gabon, despite low population and huge wealth in other natural resources, major capital inflows from the oil sector have locked the Gabonese economy into oil production.

The mining industry in the Basin features both industrial mining operators and artisanal and small-scale operators. Small-scale miners exploit deposits through rudimentary technologies and toxic chemicals to extract and process gold, tin, coltan, and diamonds. Industrial operators usually rely on mechanized equipment to access ore bodies located beneath the surface.

Promising Prospects for the Mining Sector in the Congo Basin

There are positive prospects for the development of the mining sector in the Congo Basin. Peace has been restored in most parts of the region, and many companies have returned. The rising price of many minerals worldwide drives the interest in mining companies in the Basin: reserves that used to be considered financially unviable are now receiving particular attention because of high prices and high demand. The exploding need (especially from China) for minerals largely changes the rules of game to the benefit of the Basin countries.

After 2000, the world demand for mineral resources grew significantly and reached a historic high in the middle of 2008. This increase was mainly driven by global economic development and, in particular, by China's economic growth. The pressure on demand was followed by a major rise in metal prices. Some metals saw their values triple in a short period of time. In September 2008, the global recession strongly affected the mining sector. In early 2009, aluminum and copper saw a drop in world demand by 19 percent and 11 percent, respectively. However, strong industrial development as well as renewed investments in infrastructure, construction, and manufacturing in China led to the revival of demand for raw materials[34] in the second half of 2009 (figure 2.16).

Asian countries are becoming the major importers of the mineral commodities known to exist in the Congo Basin. As of 2010, China and other Asian countries import the majority of the world's iron ore, manganese, lead, tin, aluminum, copper, cobalt, and titanium (table 2.14). Europe and the United States continue to import significant, but much smaller, quantities of titanium, cobalt, aluminum, lead, iron ore, and manganese. The exceptions to these trends are uranium ore (imported predominantly by the United States), titanium (the United States, China, Germany, and Japan account for more than half of global imports), and diamonds (imports are shared equally among the United States; Belgium; and Hong Kong SAR, China). These three commodities are used in high-end applications (power plants, aircraft, and jewelry, respectively), which traditionally have been dominated by richer economies. However, as the Chinese become wealthier, the import balance for these commodities is likely to tip in China's favor.

The decline in oil reserves also pushes countries to focus on other sectors to offset the foreseeable revenue gap. This is the case in Gabon and Cameroon, where the predicted decline in oil reserves has already started shifting economic priorities and encouraging the development of other high-value resources, such as minerals.

Figure 2.16 Commodity Metals Price Index

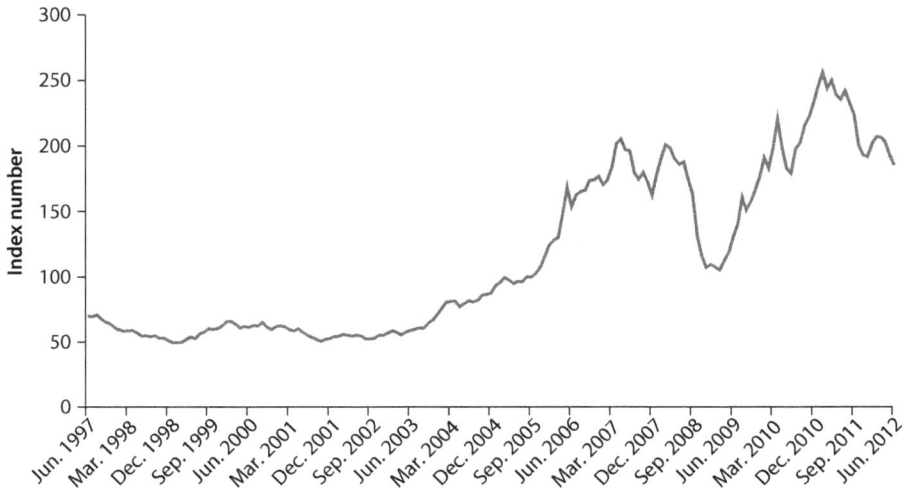

Source: Index Mundi, based on International Monetary Fund (IMF) databases (http://www.indexmundi.com/commodities/?commodity=metals-price-index&months=180; accessed July 2012).
Note: 2005 = 100, includes copper, aluminum, iron ore, tin, nickel, zinc, lead, and uranium price indices. http://www.index mundi.com/commodities/?commodity=metals-price-index&months=180.

Table 2.14 Major Importers of Mineral Commodities Known to Occur in the Congo Basin, 2010

Commodity[a]	Economy	Trade value (million US$)	Share of value (%)
Aluminum	China	4,684.28	36.9
	United States	2,046.95	17.0
	Germany	793.65	6.6
	Spain	707.23	5.9
	Ireland	604.95	5.0
Cobalt	China	2,857.62	76.0
	Finland	468.24	12.4
	Zambia	303.87	8.0
Copper	Japan	40,831.89	32.5
	China	40,266.99	32.1
	Korea, Rep.	10,154.05	8.1
	Germany	8,712.76	6.9
Diamonds[b]	United States	70,100.19	22.9
	Belgium	56,073.83	18.3
	Hong Kong SAR, China	47,906.70	15.9
	Israel	33,025.45	10.8
Iron	China	224,369.97	62.3
	Japan	46,049.68	12.8
	Germany	15,852.91	4.4
	Korea, Rep.	11,240.82	3.1
Lead	China	7,486.04	47.0
	Korea, Rep.	1,791.29	11.2
	Japan	1,409.43	8.8
	Germany	1,390.77	8.7
	Belgium	1,175.83	7.4
Manganese	China	9,347.35	58.1
	Japan	1,380.60	8.9
	Norway	1,115.36	6.9
	Korea, Rep.	718.58	4.5
Tin	Malaysia	488.88	40.7
	Thailand	435.81	38.3
	China	195.45	16.3
Titanium	United States	1,045.52	19.7
	China	743.96	14.0
	Germany	620.05	11.7
	Japan	476.20	9.0
Uranium	United States	2,479.31	98.8
	China	19.93	.8
	France	7.17	.3

Source: UN Comtrade. 2011. United Nations Statistics Division (http://comtrade.un.org; accessed November 2011).
a. Ores and concentrates, unless otherwise noted.
b. Other than sorted industrial diamonds, whether or not worked, but not mounted or set.

New mining deals, including infrastructure components, emerge. Poor infrastructure has generally been a major obstacle in the development of mining operations in the Congo Basin. However, with the high demand for minerals as well as the high prices and incentives to mine new mineral deposits, one can witness a new generation of deals. In fact, over the last few years, a trend toward investors offering to build associated infrastructures has emerged. Substantial examples include roads, railways, power plants (including large dams), and ports, among others. In Gabon, the Belinga iron ore reserves have been put under contract for development by China National Machinery and Equipment Import and Export Corporation, and this contract encompasses building the related infrastructure (railway, hydropower plant, and deep-sea ports). In Cameroon, Sundance, an Australian company, has been allotted exploration rights that would involve the development of an iron ore mine and the related infrastructure, which also falls within the dense tropical forests that cover the southern portion of the country. These new deals largely remove the burden from host countries, which generally lack the financial resources to cover the investment needs. Such deals would circumvent one of the major weaknesses of the Basin countries for the development of the mining operations.

Impacts on Forests: Current and Future
The nature of the potential impacts of mining operations on forests is varied and can range from direct, indirect, induced, and cumulative. None of them can be disregarded and need to be taken into account in order to reconcile mining development and forest wealth in the critical ecosystem of Congo Basin forests.

- Direct impacts: Direct impacts from mining entail deforestation that encompasses the following: the site covered by roads, mines, excavated minerals and earth, equipment, and settlements associated with the mining activities. Compared to other economic activities (for example, agriculture), the area deforested as a result of mining is fairly limited. However, restoring the tropical forest ecosystem is challenging and costly. Even when best practices for restoration and reclamation are used, the resulting forest ecosystem has been modified from its original, premining state. At the site itself, the degree of disturbance is a function of both the ore grade and the type of mine operation (for example, strip mine versus underground). Typically open-pit and strip-mining operations create the greatest level of land disturbance, especially in areas where the ores are deposited slightly deeper.

- Indirect impacts: Mining operations have an indirect impact on Basin forests by bringing infrastructure development to the region which, in turn, could lead to deforestation and forest degradation. Indirect impacts of mining can cover a much larger area and may include road development in the region where the mine exists and hydropower plants to supply the energy-intensive mining industry.

- Induced impacts: Mining operations are usually accompanied by a large influx of workers. They bring additional socioeconomic activities—such as subsistence agriculture, logging, and poaching—with potentially significant harm to forests.

- Cumulative impacts: In the case of artisanal mining, although individual sites may have fairly small and localized impacts on local vegetation, wildlife, and habitats, the cumulative impacts of hundreds of artisanal mining sites around the country may lead to increased risk of deforestation, habitat conversion, and biodiversity loss.

Mining activities are expected to become a major pressure on Congo Basin forests. So far, mining has had limited impacts on Basin forests because the majority of the mining operations in the Basin have occurred in nonforested areas. But with mining exploration taking off in the Congo Basin, this will also increase its impact on the forest, as described above.

Conflicting land use plans carry the potential for large-scale deforestation and forest degradation. Numerous conflicts have been noted between and among conservation priorities, mining and logging concessions, and the livelihoods of the local populations. As an example, Chupezi, Ingram, and Schure (2009) reported in the Sangha Tri-National Park region (shared between Cameroon, the Central African Republic, and the Republic of Congo) that many logging and mining concessions were overlapping with each other as well as with the region's protected areas and agroforestry zones. Poor land use management can potentially amplify the adverse impacts of mining operations, both exploration and exploitation.

Notes

1. Concept and structure of GLOBIOM are similar to the U.S. Agricultural Sector and Mitigation of Greenhouse Gas model.
2. The land cover data for 2000 is taken from the GLC2000.
3. Mosnier et al. (2012), prepared by the IIASA team, is an output of this study.
4. The Comprehensive Africa Agriculture Development Program (CAADP) was established as part of the African Union's New Partnership for Africa's Development (NEPAD) Planning and Coordinating Agency (NPCA) and was endorsed by the African Union Assembly in July 2003. CAADP's goal is to help African countries reach and sustain a higher path of economic growth through agriculture-led development that reduces hunger and poverty, and enables food and nutrition security and growth in exports through better strategic planning and increased investment in the sector. Through CAADP, African governments are committed to raising agricultural GDP by at least 6 percent per year. This is the minimum required if Africa is to achieve agriculture-led socioeconomic growth. To achieve this, these governments have agreed to increase public investment in agriculture to a minimum of 10 percent of their national budgets—substantially more than the 4 to 5 percent average they commit today.
5. Democratic Republic of Congo's "Cinq Chantiers," Republic of Congo's "Vision 2025 Pays Emergent," Cameroon's "Vision 2025," and "Gabon Emergent, 2025."
6. Because of the prevalence of tsetse fly, livestock production is marginal and limited to small ruminants, poultry, and pigs in small numbers, essentially for own use.

7. Some other crops like beans, gourds, and vegetables are also grown in home gardens together with fruit trees.

8. In some parts, upland rice is also grown as a cash crop.

9. There were a few large commercial coffee or cocoa plantations in the Democratic Republic of Congo, but almost all of them have been abandoned following first the Zairanization (expropriation) in 1973–74 and the pillages of 1991 and 1993.

10. FAOSTAT. 2011. http://faostat.fao.org/, Food and Agriculture Organization of the United Nations (FAO), Rome (accessed December 2011).

11. Food includes all agricultural commodities except the ones not used for human consumption (for example, rubber, cotton, animal feed, seed).

12. If forests are excluded, they represent about 20 percent of the land available in Sub-Saharan Africa and 9 percent at the world level.

13. Between 2000 and 2007, poultry exports increased by a factor of 23, and beef exports rose by a factor of 7. In China, the soybean imports have increased by a factor of 2.6 between 2000 and 2007 in order to augment the country's own livestock production.

14. Second-generation biofuel should also reduce the pressure on land, improving biomass energy conversion and extending usable biomass resources, but the technologies are not yet commercially available. Production of biodiesel from used cooking oil or low-grade tallow (for example, *jatropha* that can grow on some low-productivity land in Asia and Africa) is also under experiment; however, their use is currently marginal in total biodiesel production, and their potential large-scale future use is questioned (see Brittaine and Lutaladio 2010 for discussion on jatropha potential).

15. This chapter exclusively focuses on wood-based biomass energy, based on the predominant use of such energy in the Basin countries. The authors would like to highlight the fact that hydropower plants may also impact forests, as they can induce immersion of large forested areas. There are a few planned (or under implementation) investments for large-scale hydropower plants in the Congo Basin region; this aspect has, however, not been covered as part of this study.

16. In this document and in line with the definition from Miranda et al. (2010), "woodfuel" refers to both fuelwood and charcoal. Fuelwood is harvested and used directly, without further conversion. Charcoal is made from wood through the process of pyrolysis (slow heating without oxygen) and is typically used by households or small- and medium-sized businesses.

17. A study for Dar es Salaam suggests that a 1 percent increase in urbanization leads to a 14 percent rise in charcoal consumption (Hosier, Mwandosya, and Luhanga 1993).

18. The exact numbers are difficult to estimate for this mainly informal sector for which no data are available, but people depending on this sector for their livelihood tend to be members of poorer households (who work as small-scale producers/collectors, traders, transporters, or retailers), often with limited alternatives for earning cash income.

19. All the data used in this section are based on the Africa Infrastructure Country Diagnostic (AICD) reports (Foster and Briceño-Garmendia 2010) and database (http://www.infrastructureafrica.org/tools; accessed March 2012). The AICD, spearheaded by the World Bank, has provided a comprehensive assessment of the needs

for physical infrastructure (as well as associated costs) in Sub-Saharan Africa. It has collected detailed economic and technical data on each of the main infrastructure sectors, including energy, information and communication technologies, irrigation, transport, and water and sanitation.

20. In reality, the good performance in the Central African Republic hides the fact that the vast majority of the classified roads are paved roads, which only represent one-third of the total roads, and only 2 percent of the classified unpaved roads meet the standard of good and fair conditions.

21. The Road Transport Quality index is calculated from a formula combining the following parameters: Q = Road quality index for a country; P = Percentage of roads that are paved in a country; G = GDP per capita in a country (an index of capacity to maintain roads); and C = The World Bank's Country Policy and Institutional Capacity index for transparency, accountability, and corruption in a country (a proxy for delays and costs inflicted on truckers).

22. The Central African Republic does not have a single all-season road corridor to its coastal port gateways.

23. CEMAC, Agreed Steering Program for Transport in Central Africa (Plan Directeur Consensuel pour le Transport), ECCAS 2004.

24. Some of the many studies on the topic are Cropper, Griffiths, and Mani (1999); Pfaff (1999); Chomitz et al. (2007); and Soares-Filho et al. (2005). Studies looking specifically to the Congo Basin include Mertens and Lambin (1997), Wilkie et al. (2000), and Zhang et al. (2002, 2005, and 2006).

25. Indirect impacts are usually of greater magnitude along a road than a railway.

26. The authors are aware of the limitations of such an assumption, as the literature presents various examples where this direct correlation between time and costs does not apply. However, in the absence of a stronger assumption, this one has been applied.

27. Consumers in the Congo Basin rely more and more on imported agricultural products. A reduction in "prices to consumer" can favor the locally grown products.

28. Without any changes in the other parameters.

29. OFAC. National Indicators. 2011. www.observatoire-comifac.net, Kinshasa (accessed March 2012).

30. Note that data on sawmills varies widely across the literature and not all sawmills listed are actually operational and active. In particular, in the Democratic Republic of Congo, sawmills are abandoned and inactive after many years of neglect during the years of civil conflict.

31. This SEZ will be dedicated to advanced processing of tropical timber, with a global capacity of 1 million cubic meters per year and direct employment estimated between 6,000 and 7,000. As of November 2011, US$200 million have already been invested in this joint venture SEZ. The SEZ is expected to be operational in mid-2012.

32. This feature was the main reason for the Basin countries to join forces during the Conference of Parties in Bali in 2007 and expand the concept of RED to forest degradation (thus adding the second "D" to the acronym REDD).

33. In this section, we consider the impacts in terms of carbon content (as per the REDD+ mechanism). However, it is important to note that although logging may

have limited impacts on carbon stock on the long run, impacts on biodiversity and ecosystem equilibrium may be much more affected by logging.

34. When a country holds a substantial portion of the production and consumption of a good, like in the case of China today, domestic events can significantly impact the world price of the commodity.

References

Angelsen, A., M. Brockhaus, M. Kanninen, E. Sills, W. D. Sunderlin, and A. Wertz-Kanounnikoff. 2009. *Realising REDD+: National Strategy and Policy Options.* Bogor, Indonesia: Center for International Forestry Research (CIFOR).

AICD. 2012. *Africa Infrastructure Country Diagnostic database.* Accessed March 2012. http://www.infrastructureafrica.org/tools.

Blaser, J., A. Sarre, D. Poore, and S. Johnson. 2011. *Status of Tropical Forest Management 2011.* Technical Series 38. Yokohama, Japan: International Tropical Timber Organization.

Brittaine, R., and N. Lutaladio. 2010. *Jatropha: A Smallholder Bioenergy Crop. The Potential for Pro-Poor Development.* Rome, Italy: Integrated Crop Management Food and Agriculture Organization of the United Nations.

Brown, S., T. Pearson, N. Moore, A. Parveen, S. Ambagis, and D. Shoch. 2005. *Impact of Selective Logging on the Carbon Stocks of Tropical Forests: Republic of Congo as a Case Study.* Arlington, VA: Winrock International.

Bruinsma, J. 2009. "The Resource Outlook for 2050: By How Much Do Land, Water and Crop Yields Need to Increase by 2050?" Paper presented at the Expert Meeting on How to Feed the World in 2050, "Food and Agriculture Organization of the United Nations," Rome, Italy.

CEMAC (Economic and Monetary Community of Central Africa). 2004. "Agreed Steering Program for Transport in Central Africa." Bangui, Central African Republic.

———. 2009. *CEMAC 2025: Towards an Integrated Emerging Regional Economy: Regional Economic Program 2010–2015 (Vers une économie régionale intégrée et émergente Programme Economique Régional 2010–2015).* Volume 2. Bangui, Central African Republic.

Cerutti, P. O., and G. Lescuyer. 2011. "The Domestic Market for Small-Scale Chainsaw Milling in Cameroon: Present Situation, Opportunities and Challenges." Occasional Paper 61, Center for International Forestry Research (CIFOR), Bogor, Indonesia.

Chomitz, K. M., P. Buys, G. De Luca, T. S. Thomas, and S. Wertz-Kanounnikoff. 2007. *At Loggerheads? Agricultural Expansion, Poverty Reduction, and Environment in the Tropical Forests.* Washington, DC: World Bank.

Chupezi, T. J., V. Ingram, and J. Schure. 2009. *Study on Artisanal Gold and Diamond Mining on Livelihoods and the Environment in the Sangha Tri-National Park Landscape, Congo Basin.* Yaounde, Cameroon: Center for International Forestry Research (CIFOR)/International Union for Conservation of Nature (IUCN).

Collier, P. 2007. *The Bottom Billion, Why the Poorest Countries Are Failing and What Can Be Done About It.* Oxford: Oxford University Press.

Cropper, M., C. Griffiths, and M. Mani. 1999. "Roads, Population Pressures and Deforestation in Thailand, 1976–1989." *Land Economics* 75 (1): 58–73.

Deininger, K., D. Byerlee, J. Lindsay, A. Norton, H. Selod, and M. Stickler. 2011. *Rising Global Interest in Farmland—Can It Yield Sustainable and Equitable Benefits?* Washington, DC: World Bank.

Duveiller, G., P. Defourny, B. Desclée, and P. Mayaux. 2008. "Deforestation in Central Africa: Estimates at Regional, National and Landscape Levels by Advanced Processing of Systematically-Distributed Landsat Extracts." *Remote Sensing of Environment* 112 (5): 1969–81.

FAO (Food and Agriculture Organization of the United Nations). 2009a. *The State of Food Insecurity in the World.* Rome: FAO.

———. 2009b. *Statistical Yearbook 2009.* Rome: FAO.

Foster, V., and C. Briceño-Garmendia, eds. 2010. *Africa Infrastructure Country Diagnostic: A Time for Transformation.* Main report and associated background papers and working papers. Washington, DC: World Bank.

Geist, H., and E. Lambin. 2001. "What Drives Tropical Deforestation: A Meta-Analysis of Proximate and Underlying Causes of Deforestation Based on Subnational Case Study Evidence." Land-Use Land-Cover Change Report Series 4, LUCC International Project Office, Louvain-la-Neuve, Belgium.

———. 2002. "Proximate Causes and Underlying Driving Forces of Tropical Deforestation." *BioScience* 52 (2): 143–50.

Gibbs, H., A. Ruesch, F. Achard, K. Clayton, P. Holmgren, N. Ramankutty, and A. Foley. 2010. "Tropical Forests Were the Primary Sources of New Agricultural Land in the 1980s and 1990s." *Proceedings of the National Academy of Sciences* 107 (38): 16732–37.

Hiemstra-van der Horst, G., and A. J. Hovorka. 2009. "Woodfuel: The 'Other' Renewable Energy Source for Africa?" *Biomass and Bioenergy* 33 (11): 1605–16.

Hosier, R. H., M. J. Mwandosya, and M. L. Luhanga. 1993. "Future Energy Development in Tanzania: The Energy Costs of Urbanization." *Energy Policy* 35 (8): 4221–34.

Howitt, R. E. 1995. "Positive Mathematical Programming." *American Journal of Agricultural Economics* 77 (2): 23–31.

Hoyle, D., and D. Levang. 2012. "Oil Palm Development in Cameroon." Ad Hoc Working Paper, World Wildlife Fund in partnership with Institut de Rechercherund pour le Développement (IRD) and Center for International Forestry Research (CIFOR), Bogor, Indonesia.

IEA (International Energy Agency). 2006. *World Energy Outlook 2006.* Paris, France: Organisation for Economic Co-operation and Development/IEA.

———. 2010. *World Energy Outlook 2010.* Paris, France: Organisation for Economic Co-operation and Development/IEA.

IIASA (International Institute for Applied Systems Analysis). 2011. *Modeling Impacts of Development Trajectories on Forest Cover and GHG Emissions in the Congo Basin.* Washington, DC: World Bank.

ITTO (International Tropical Timber Organization). 2006. *Status of Tropical Forest Management 2005.* Yokohama, Japan: ITTO.

Langbour, P., J.-M. Roda, and Y. A. Koff. 2010. "Chainsaw Milling in Cameroon: The Northern Trail." *European Tropical Forest Research Network News* 52: 129–37.

Laurance, W., M. Goosem, and S. Laurance. 2009. "Impacts of Roads and Linear Clearing on Tropical Forests." *Trends in Ecology & Evolution* 24 (12): 659–69.

Leach, G. 1992. "The Energy Transition." *Energy Policy* 20 (2): 116–23.

Lescuyer, G., P. O. Cerutti, E. E. Mendoula, R. Ebaa-Atyi, and R. Nasi, 2010a. "Chainsaw Milling in the Congo Basin." *European Tropical Forest Research Network News* 52: 121–28.

Lescuyer, G., P. O. Cerutti, S. N. Manguiengha, and L. B. bi Ndong. 2010b. "The Domestic Market for Small-Scale Chainsaw Milling in Gabon: Present Situation, Opportunities and Challenges." Occasional Paper 65, CIFOR, Bogor, Indonesia.

Lescuyer, G., P. O. Cerutti, E. E. Mendoula, R. Eba'a Atyi, and R. Nasi. 2012. "An Appraisal of Chainsaw Milling in the Congo Basin." In *The Forests of the Congo Basin—State of the Forest 2010*, ed. de Wasseige et al. Luxembourg: Publications Office of the European Union.

Marien, J. N. 2009. "Peri-Urban Forests and Wood Energy: What Are the Perspectives for Central Africa?" In *The Forests of the Congo Basin—State of the Forest 2008*, ed. de Wasseige et al. Luxembourg: Publications Office of the European Union.

McCarl, B., and T. Spreen. 1980. "Price Endogenous Mathematical Programming as a Tool for Sector Analysis." *American Journal of Agricultural Economics* 62 (1): 87–102.

Mertens, B., and E. Lambin. 1997. "Spatial Modeling of Deforestation in Southern Cameroon: Spatial Disaggregation of Diverse Deforestation Processes." *Applied Geography* 17 (2): 143–62.

Miranda, R., S. Sepp, E. Ceccon, S. Mann, and B. Singh. 2010. *Sustainable Production of Commercial Woodfuel: Lessons and Guidance from Two Strategies*. Washington, DC: Energy Sector Management Assistance Program (ESMAP), World Bank.

Mosnier, A., P. Havlik, M. Obersteiner, and K. Aoki. 2012. *Modeling Impacts of Development Trajectories on Forest Cover in the Congo Basin*. Laxenburg, Austria: International Institute for Applied Systems Analysis Environment Research and Education (ERE).

Nasi, R., P. Mayaux, D. Devers, N. Bayol, R. Eba'a Atyi, A. Mugnier, B. Cassagne, A. Billand, and D. Sonwa. 2009. "A First Look at Carbon Stocks and Their Variation in Congo Basin Forests." In *The Forests of the Congo Basin—State of the Forest 2008*, ed. de Wasseige et al. Luxembourg: Publications Office of the European Union.

Pfaff, A. 1999. "What Drives Deforestation in the Brazilian Amazon? Evidence from Satellite and Socio-Economic Data." *Journal of Environmental Economics and Management* 37 (1): 26–43.

Reed, E., and M. Miranda. 2007. *Assessment of the Mining Sector and Infrastructure Development in the Congo Basin Region*. Washington, DC: World Wildlife Fund.

SEI (Stockholm Environment Institute). 2002. *Charcoal Potential in Southern Africa, CHAPOSA: Final Report*. Stockholm: INCO-DEV, Stockholm Environment Institute.

Soares-Filho, B., A. Alencar, D. Nepstad, G. Cerqueira, M. Vera Diaz, S. Rivero, L. Solorzano, and E. Voll. 2005. Simulating the Response of Land-Cover Changes to Road Paving and Governance along a Major Amazon Highway: The Santarém–Cuiabá Corridor." *Global Change Biology* 10 (5): 746–64.

Stickler, C, M. Coe, D. Nepstad, G. Fiske, and P. Lefebvre. 2007. *Readiness for REDD: A Preliminary Global Assessment of Tropical Forested Land Suitability for Agriculture*. Falmouth, MA: The Woods Hole Research Center.

Teravaninthorn, S., and G. Raballand. 2008. "Transport Prices and Costs in Africa: A Review of the Main International Corridors." AICD Working Paper 14, World Bank, Washington, DC.

Tollens, E. 2010. *Potential Impacts of Agriculture Development on the Forest Cover in the Congo Basin*. Washington, DC: World Bank.

Trefon, T., T. Hendriks, N. Kabuyaya, and B. Ngoy, 2010. "L'économie Politique de la Filière du Charbon de Bois à Kinshasa et à Lubumbashi." Working Paper 2010.03, Universiteit Antwerpen, Institute of Development Policy and Management (IOB), Antwerp, Belgium.

UN-Energy Statistics (database). http://data.un.org/. United Nations Statistics Division. Consulted in November 2011.

Wilkie, D., E. Shaw, F. Rotberg, G. Morelli, and P. Auzel. 2000. "Roads, Development and Conservation in the Congo Basin." *Conservation Biology* 14 (6): 1614–22.

Williams, J. 1995. *The EPIC Model. Computer Models of Watershed Hydrology*, 909–1000. Highlands Ranch, CO: Water Resources Publications.

World Bank. 2009. *Environmental Crisis or Sustainable Development Opportunity? Transforming the Charcoal Sector in Tanzania. A Policy Note*. Washington, DC: World Bank.

———. 2010. *Doing Business 2010*. Washington, DC: World Bank. http://www .doingbusiness.org/.

———. 2012. *Household Energy for Cooking and Heating: Lessons Learned and Way Forward*. Washington, DC: World Bank.

Zhang, Q., D. Devers, A. Desch, C. Justice, and J. Townshend. 2005. "Mapping Tropical Deforestation in Central Africa." *Environmental Monitoring and Assessment* 101 (1–3): 69–83.

Zhang, Q., C. Justice, P. Desanker, and J. Townshend. 2002. "Impacts of Simulated Shifting Cultivation on Deforestation and the Carbon Stocks of the Forests of Central Africa." *Agriculture, Ecosystems & Environment* 90 (2): 203–9.

Zhang, Q., C. Justice, M. Jiang, J. Brunner, and D. Wilkie. 2006. "A GIS-Based Assessment on the Vulnerability and Future Extent of the Tropical Forests of Congo Basin." *Environmental Monitoring and Assessment* 114 (1–3): 107–21.

CHAPTER 3

REDD+: Toward a "Forest-Friendly" Development in the Congo Basin

This chapter explores the benefits and opportunities that the new mechanism reducing emissions from deforestation and forest degradation plus (REDD+; under discussion in the Climate Change negotiations) affords to Congo Basin countries. The first section provides an overview of REDD+ and relevant concepts, along with the potential financing and related challenges Basin countries should address to secure such financing. The second section offers an analysis of policies and actions that could help Basin countries reconcile their urgent need to develop economically and to preserve their natural forests. It highlights recommendations that can serve as general guidelines and could spur more detailed policy discussions at a national level as countries move toward preparing their REDD+ strategies. Recommendations are divided into crosscutting issues and enabling dimensions that affect multiple sectors—land use planning, land tenure, and law enforcement—and sector-specific actions for agriculture, energy, transport, logging, and mining sectors.

REDD+: A New Mechanism to Reduce Pressure on Tropical Forests

Forests: An Integral Part of Climate Change Negotiations

In 2005, the Parties of United Nations Framework Convention on Climate Change (UNFCCC) initiated the debate to include tropical deforestation within climate change discussions. In 1997, the parties had decided not to include the tropical deforestation in the Protocol of Kyoto, although it was already clear that the issue largely contributes to the global greenhouse gas (GHG) emissions. In 2005, at the Montreal Conference, Papua New Guinea and Costa Rica relaunched the debate with a joint submission entitled "Avoided Deforestation." Since then, UNFCCC parties have had extensive discussions about the scope of REDD+, and in 2009, in Copenhagen, they reached a consensus on the REDD+ concept (it was then confirmed during the following Conference of Parties in Cancun and Durban).

In 2007, the UNFCCC parties first embraced the concept of compensating developing nations in order to slow deforestation and thus reduce carbon emissions in the atmosphere. The idea of setting up an incentive mechanism, known as REDD+ mechanism, to reduce deforestation in developing countries has been debated and increasingly substantiated in the past five years. The concept of the incentive mechanism, which initially focused on reducing emissions from deforestation (RED), has been expanded and now encompasses the reduction in forest degradation as well as the promotion of conservation, sustainable forest management (SFM), and enhancement of carbon stocks, and is referred to as REDD+ (reducing emissions from deforestation and forest degradation plus). REDD+ is now likely to be one of the prominent features of the post-2012 Climate regime (that is, it is now enshrined in the International Agreement adopted during the Conference of Parties [COP-16] in Cancun in November 2010). For more details on the various concepts, see box 3.1.

REDD+ is now firmly accepted by all countries and will be embedded in future climate regimes. As currently defined, the concept embraces "reducing emissions from deforestation and forest degradation, and the role of conservation, sustainable management of forests, and enhancement of forest carbon stocks." More specifically, many countries expect continuing UNFCCC discussions to result in a mechanism that will help finance the shared goal of "slowing, halting, and reversing forest cover and carbon loss" through results-based incentives for

---

Box 3.1 The Evolving Scope of REDD+ under International Discussions

Since the start of negotiations on deforestation, various concepts have been debated:

- RED (reduction of emissions from deforestation), the early stage concept, would be restrained to the lands that are switched from "forest" to another kind of use (nonforest). It would therefore not encompass logging activities if the number of logged trees was insufficient to change the land use according to the national definition of what constitutes a forest—as in Central Africa.
- REDD (reduction of emissions from deforestation and forest degradation), which captures both the deforestation and the carbon stock density, decreases within the forest. In 2007, in Bali, the strong involvement of the Congo Basin countries was seen as a major contribution to the inclusion of forest degradation in the REDD mechanism.
- REDD+ (or REDD-plus) covers REDD activities, but it also considers the role of conservation, sustainable forest management (SFM), and enhancement of forest carbon stock in developing countries. Therefore, in addition to the avoided emissions, REDD+ is also examining the sequestration role of forest (carbon sink). It also highlights the function of SFM, as compared to regular logging, in reducing emissions.
- REDD++, as a first move toward a broader agriculture, forest, and other land uses system, is taking into account emissions from agriculture and other land uses. Although REDD+ is limited to "forestlands," REDD++ can encompass other carbon stocks—such as agroforestry and trees outside forest—and does not depend on the operational definition of "forest."

---

measured, reported, and verified emission reductions from forest protection. This expanded definition of eligible activities in the context of UNFCCC negotiations has created potential opportunities for forest protection in the Congo Basin.

That said, the concept of REDD+ keeps evolving, and a mechanism that provides direct payments for verified emission reductions is far from settled. Outstanding uncertainties continue to pose challenges for developing countries, meaning that REDD+ and its implementation in the Basin is not without unknowns and risks. In this context, several relevant issues, opportunities, and challenges are outlined below.

A three-phase approach was adopted for the REDD+ mechanism. The Cancun (COP-16) UNFCCC decision on REDD+[1] establishes that REDD+ activities undertaken by developing countries should be implemented in phases: first, the development of national strategies and capacity building; second, the implementation of national action plans that could involve further capacity building, technology development and transfer, and results-based demonstration activities; and third, results-based actions that should be fully measured, reported, and verified (figure 3.1). It is, however, important to keep in mind that three phases are not purely sequential and that there can be overlap between them, as is already observed in many countries where readiness activities overlap with demonstration projects.

Financing Opportunities Related to REDD+

Increased international attention to climate change has resulted in new funds allocated for REDD+ activities. In December 2009, at COP-15 in Copenhagen, parties agreed in the (informal) Copenhagen Accord that "scaled-up, new and additional, predictable, and adequate funding" shall be provided to developing countries "to enable and support enhanced action on mitigation, including substantial financing to reduce emissions from deforestation and forest degradation (REDD+)." The Accord also put forth a "collective commitment by developed countries to provide new and additional resources, including forestry and investments through international institutions, approaching

Figure 3.1 Three-Phase Approach of the REDD+ Mechanism

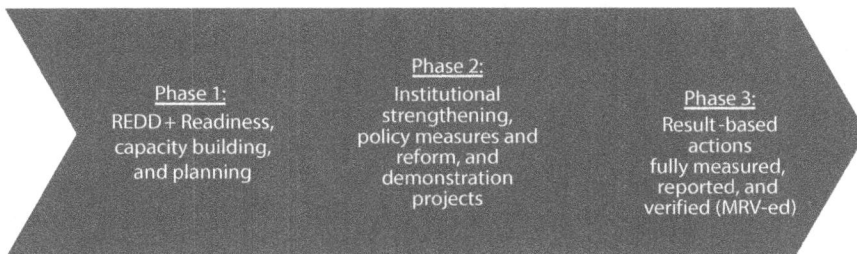

Phase 1:
REDD+ Readiness, capacity building, and planning

Phase 2:
Institutional strengthening, policy measures and reform, and demonstration projects

Phase 3:
Result-based actions fully measured, reported, and verified (MRV-ed)

US$30 billion for the period 2010–2012," which is generally referred to as fast start financing.

The majority of "fast start" REDD+ funds[2] are disbursed through bilateral donors. In this initial phase of climate finance, Norway has been a prominent REDD+ donor. As early as 2007, at COP-13 in Bali, the government of Norway launched its International Climate and Forest Initiative (NICFI) and pledged US$2.5 billion in new and additional funds for cost-effective and verifiable reductions in GHG emissions from deforestation. Since then, Norway has entered into bilateral agreements with Brazil, Indonesia, Guyana, Ethiopia, and Tanzania and contributed to various multilateral funds. As part of its bilateral agreements, Norway has pursued a "payment for performance" approach toward REDD+. Other major donors to REDD+ activities include Australia, France, Germany, Japan, the United Kingdom, and the United States. Until now, these donors have mostly supported readiness programs, policy support, and demonstration projects.

A substantial share of the current REDD+ funding has also been directed to multilateral funds and programs. These funds include the Forest Carbon Partnership Facility (FCPF) and Forest Investment Program (FIP), both of which are trust funds of the World Bank, UN-REDD Program, and the Global Environment Facility (GEF). Such programs support "readiness," implementation of policy measures and reform as well as pilots for results-based funding. Some funds like the GEF trust fund have REDD+ as one of the categories of mitigation activities funded, while other funds like the FCPF exclusively fund REDD+ initiatives. Table 3.1 summarizes the most important multilateral funds that contribute to REDD+.

The three phases of the REDD+ mechanism have often been linked to particular funding sources (table 3.2). In phase 1, the so-called readiness phase, countries build REDD+ strategies, begin to create monitoring systems, and address social and environmental safeguards—activities that are expected to be mainly supported by public funds. Moving to the implementation of policies and measures that address deforestation drivers requires a more sustained and reliable flow of international GEF or national finance (phase 2). Finally, many hope that phase 3 will reward verifiable success of GHG emission reduction through the provision of performance-based payments. Financing for this final stage, however, has not yet been clarified under the UNFCCC, neither has any country to date passed legislation or created a system that would regulate such transactions.

Fast start financing is seen only as a beginning, and many expect funding for climate change to increase over time. At COP-16 in Cancun, a new set of decisions were adopted. They state that "developed country Parties commit, in the context of meaningful mitigation actions and transparency on implementation, to a goal of mobilizing jointly US$100 billion per year by 2020 to address the needs of developing countries." This scaled up funding is understood to be a combination of both public funds and the mobilization of private-sector investments.

Table 3.1 Multilateral Reducing Emissions from Deforestation and Forest Degradation Plus (REDD+) Funds

Fund	Description	Geographic focus	Pledged January 2012 (US$ million)
The Forest Carbon Partnership Facility (FCPF) Readiness Fund	World Bank trust fund launched in 2007 for REDD+ capacity building. The FCPF Readiness Fund is currently supporting 36 developing countries to build REDD+ strategies, MRV systems, and national baselines; includes knowledge sharing among members.	13 countries in Africa, 15 in Latin America, and 8 in Asia-Pacific	229.4
FCPF Carbon Fund	Partner to the Readiness Fund; declared operational in 2011; will provide performance-based payments for verified emission reductions from REDD+; only countries that have achieved progress toward REDD+ readiness will be eligible.	Only countries from FCPF Readiness Fund are currently eligible	204.4
UN-REDD+	Collaboration between the Forest and Agriculture Organization, the UN Development Program, and the UN Environment Program to support the development of national REDD+ readiness.	Bolivia, Democratic Republic of Congo, Ecuador, Indonesia, Panama, Papua New Guinea, Tanzania, Vietnam, Zambia	151
Forest Investment Program (FIP)	World Bank Climate Investment Fund operational since July 2009. Supports "phase 2" of REDD+ activities and designed to provide scaled-up financing for forest-sector reforms identified through national REDD+ strategies.	Brazil, Burkina Faso, Democratic Republic of Congo, Ghana, Indonesia, Mexico, Lao PDR, Peru	578
Global Environmental Facility (GEF)— Climate Change Focal Area	Financial mechanism for the United Nations Framework Convention on Climate Change (UNFCCC), the UN Convention on Biological Diversity (UNCBD), and the UN Convention to Combat Desertification (UNCCD); supports projects that benefit the global environment and promote sustainable livelihoods. Various activities include the development of small-scale REDD+ projects and capacity building.	Global	246.23
Congo Basin Forest Fund	Created to complement existing activities; support transformative and innovative proposals that develop the capacity of the people and institutions of the Congo Basin and enable them to manage their own forests; help local communities find livelihoods consistent with the conservation of forests; and reduce deforestation.	Central African (COMIFAC) nations	165
Green Climate Fund	Agreed under the Copenhagen Accord and Cancun Agreements; still under negotiation; to fund adaptation and mitigation (and likely to include REDD+); to be governed by "balanced representation"; the World Bank is to act as initial trustee; not yet operational.	All developing countries	Tbd

Source: Based on data from Gledhill et al. 2011 and REDD+ Partnership Voluntary REDD+ database (http://www.reddplusdatabase.org), with some updates.
Note: COMIFAC = Forestry Commission of Central Africa.

Table 3.2 The Three Phases of Reducing Emissions from Deforestation and Forest Degradation Plus (REDD+)

Phase	Activities	Principal funding sources
Phase 1	REDD+ readiness capacity building and planning	Public funds largely channeled through bilateral agencies and multilateral funds and programs
Phase 2	Institutional strengthening, policy measures and reform, and demonstration projects	Public funds through bilateral agreements, some multilateral funds, and some private finance, with public support
Phase 3	Results-based actions, fully measured, reported, and verified (MRV-ed)	*To be determined.* Likely a variety of sources, including public funds through bilateral agreements and, potentially, the Green Climate Fund; also potential private investment and carbon markets

Source: Gledhill et al. 2011.

With respect to financing for REDD+, however, it remains unclear what mechanisms might be created and what modalities would make a country eligible to access such funding streams in the future. At COP-17 in Durban, a decision on REDD+ was adopted, which recognized "that results-based financing provided to developing country Parties … may come from a variety of sources, public and private, bilateral and multilateral, including alternative sources" and that "appropriate market-based approaches could be developed by the Conference of the Parties to support results-based actions by developing country Parties." Many saw this decision as a positive step toward a pay-for-performance system, particularly for results-based actions (phase 3), in addition to promoting continued public support for phases 1 and 2, that is, strategy development, capacity building, technological progress and transfer, and demonstration activities.

The scope, role, and modalities of future results-based (phase 3), or market, mechanisms remain undefined. Discussions under the UNFCCC will need to progress in order for countries—such as those in the Congo Basin—to understand their implications. It is worth noting that even if the UNFCCC imposes rules for REDD+ crediting, developed countries would still need to create demand for such credits by developing complementary policies, such as allowing offsets into the European Union Emissions Trading System (EU ETS). Thus far, only a few schemes have started to emerge with potential inclusion of REDD+ credits. Japan has stated that it is developing bilateral offset mechanisms, including possibly REDD+, outside the Kyoto Protocol's Clean Development Mechanism; the State of California also has provisions in its cap-and-trade system that may allow forest carbon credits to enter its system as early as 2015. In addition, there is a small voluntary market for REDD+ credits.

At the same time, countries increasingly view REDD+ finance more broadly than results-based and carbon market mechanisms. In particular, countries are starting to move from a strict consideration of REDD+ as a carbon finance mechanism to a broader development agenda that would be more "forest friendly." Some countries are showing interest in engaging, on a longer term, with the private sector to foster investments in sustainable timber products, energy sources, or agricultural supply chains—particularly where these sectors are the drivers of deforestation.

Opportunities and Challenges for the Congo Basin Countries

The trajectory of international negotiations on forests and climate change has been positive for Congo Basin countries. When forests were first put on the agenda of international negotiations at the 11th session of the conference of the Parties in Montreal in 2005, the focus was limited to mitigating emissions from deforestation (see box 3.1). Since that time, the concept has expanded to include forest degradation, conservation, SFM, and enhancement of forest carbon stocks to also implicitly value standing forests. The notion of including forest degradation and conservation was largely added to the agenda due to support from the Basin countries.

REDD+ financing has already benefitted the Congo Basin countries. All six of them have committed to the REDD+ strategy and, in return, have received support from bilateral donors and/or multilateral programs through World Bank, UN funds, and the Congo Basin Forest Fund (supported by Norway and the United Kingdom, and implemented by the African Development Bank). While disbursement has been a challenge, funds remain committed through these channels.

However, challenges remain for the Congo Basin to fully access the financing opportunities under the REDD+ mechanism. The current financing resources that flow to the Basin countries fall under phase 1, which addresses the readiness process (including capacity building and planning). Transitioning to the financing mechanisms planned under phase 3 will require the Congo Basin to overcome a series of technical and methodological challenges. The result-based nature of phase 3 establishes a minimum capacity to enforce plans, measure, and monitor carbon stocks in order for countries to be rewarded. Payment will be made for performance against a baseline (or "business as usual" without REDD+) called the reference level.

One of the most critical methodological challenges for Basin countries relates to the definition of reference levels. One of the core requirements for measured, reported, and verified financial results (phase 3) is the baseline against which to measure such performance. Called "reference emissions levels" or "reference levels," the modalities upon which countries might construct these levels are currently under discussion in the UNFCCC[3] (Angelsen et al. 2011a, 2011b). How the reference levels will be defined will hugely impact the future REDD+ mechanism and the potential benefits drawn from countries. It is expected that this reference level—beyond technical considerations that help to establish a baseline—will also likely be based on negotiations. For high forest cover, low deforestation (HFLD) countries (with an HFLD profile, as is the case for the Congo Basin countries), historic baselines may not capture the effort it takes a country to combat future risks that forests face (Martinet, Megevand, and Streck 2009).

New methods for setting reference levels are emerging. Under the Kyoto Protocol, developed countries are now allowed to set reference levels based on expected future rates of deforestation. In December 2008, at COP-14 in Poznan, countries agreed that REDD+ reference levels should "take into account historic data and adjust for national circumstances." This statement appears to suggest that countries, such as those in the Congo Basin, with low historic rates of deforestation—but potentially high expected future rates—could consider factoring these adjustments to national circumstances into a proposed reference level. The application of a development adjustment factor to reflect national circumstances is one of the more controversial elements of reference levels (see box 3.2).

The challenge, however, for countries that intend to go beyond historic data is their need for credible data and a strong justification for "adjusting for national circumstances." Many experts would suggest few developing countries have adequate and credible information that would justify using an historic trend. In particular, a lack of reliable databases usually makes it difficult to project the future impact that economic growth of the various sectors would have on forests. The CongoBIOM model is an attempt for the Basin to try to use minimum data available in order to offer an initial estimate of future trends; however, currently limited data (both in terms of quantity and quality) are unlikely to provide sufficiently robust quantitative information for reference-level setting in a financing mechanism.

Some flexibility, however, is recognized for countries with low capacity and poor data. The decision in December 2011 in Durban at COP-17, "invites Parties to submit information and rationale on the development of their forest reference emission levels and/or reference levels, including details of national circumstances, and, if adjusted, include details on how the national circumstances were considered." It is worth noting that the decision also allows countries to take a step-wise approach to setting their national reference (emissions) level and permits updates of baselines over time, incorporating better data, improved methodologies, and new knowledge of trends.

Box 3.2 Controversy on Modalities to Use Adjustment Factor in the Definition of Reference Levels

When it comes to setting reference levels, there is little agreement on a methodology that satisfies all parties. The Congo Basin countries have generally supported the introduction of development adjustment factors to account for their historically low deforestation rates and to forecast higher rates. The inherent risk is the creation of a "lemons market," where the seller knows more about the quantity of the product than does the buyer. Artificially inflated reference levels could lead to an oversupply of cheap REDD+ credits. Developing fact-based development adjustment factors, as proposed by the Congo Basin countries, will ensure that REDD+ efforts are compensated in a fair manner among REDD+ countries and will guarantee the environmental integrity of REDD+ by preventing inflated emission reduction claims.

REDD+: How Can Economic Development and Forest Preservation Be Reconciled? Some Recommendations

Many developing countries are increasingly recognizing the importance of integrating REDD+ into broader, economy-wide, low-emission development strategies that provide an opportunity to reconcile their economic development and the preservation of the forests (as a local, national, and international public good). Although there are many unknowns and challenges, the benefits of pursuing REDD+ likely outweigh these risks. Many elements within a country's REDD+ strategy could build on "no-regrets" measures: these measures could contribute to economic growth while protecting ecosystems, watersheds, and natural forests, and while also consulting with and engaging local communities, regardless of the shape of a future mechanism under the UNFCCC.

This section offers recommendations that can serve as general guidelines as the Congo Basin countries engage in the preparation of their REDD+ strategy. As there are several outstanding uncertainties in the development of a future REDD+ incentives mechanism, the section focuses on "no-regrets" recommendations. These measures are necessary conditions for attracting internal funding for REDD+, but they are also expected to generate economic benefits even in the absence of such international funding or before such funding becomes available on a larger scale. Recommendations are divided into crosscutting issues and enabling dimensions that affect multiple sectors, like land use planning, land tenure, and law enforcement, and then sector-specific actions for the targeted sectors (agriculture, energy, transport, logging, and mining sectors).

Invest in Participatory Land Use Planning

Participatory land use planning should be employed to maximize economic and environmental objectives. Congo Basin countries lack a comprehensive land use plan that leads to the problems of overlapping usage titles and potentially conflicting land uses. Many conflicts have been noted between and among conservation priorities, mining and logging concessions, and the livelihoods of the local populations. A comprehensive land use planning exercise, to be conducted in a participatory manner, should determine the different land uses to be pursued on the national territories. Once completed, this land plan would establish the forest areas that need to be preserved and those that could potentially be converted into other uses. While planning for economic development, particular attention should be given to "high-value forests" in terms of biodiversity, watershed, and cultural values.

Optimally, mining, agriculture, and other activities could be directed away from forests of great ecological value. Cleary demarcated areas for mining, logging, agriculture, and woodfuel production can help maximize productivity of these areas by stimulating intensification of lands currently used for these purposes. As far as agriculture is concerned, new development should primarily target degraded lands.[4] In the Congo Basin, large amounts of nonforested lands with high potential in low-population density areas suggest that there is no need, in principle, to draw on currently forested areas to satisfy the future demand for agricultural commodities (see Agriculture section below). However, past trends show that forested areas may be more vulnerable to agriculture expansion, so if forests are to be protected, proactive measures need to be created by the governments. Priority should be given to these available and suitable nonforested lands, including the degraded areas and abandoned commercial plantations. This strategy should be one of the guiding principles when a country conducts its land use planning exercise.

Among and within different sectors, trade-offs need to be clearly understood by stakeholders so that they can define robust development strategies at the national level. This requires tough socioeconomic analysis as well as a strong coordination between the different line ministries in order to support potentially difficult arbitrage among the various priorities. High-level political involvement is usually required to eventually reconcile potential conflicting uses of lands.

One output of such an exercise could be identifying growth poles and major development corridors that could be expanded in a coordinated manner, with the involvement of all governmental entities along with private sector and civil society. A growth pole approach is used to define coordinated interventions in selected areas with the maximum potential impact in terms of economic growth. It is often characterized by a key industry around which linked industries grow, mainly through direct and indirect effects. In the Congo Basin, such an approach will likely be driven by natural resources (predominantly minerals). Resource corridors are indeed seen as a natural way to promote growth poles, as they leverage infrastructure as well as provide upstream and downstream linkages around extractive industries.

Regional dimension should be taken into account in the land use planning exercises. While such exercises definitely need to be conducted at the country level (and even provincial level) to define country-specific priorities in line with national strategies, the benefit of regional integration is also undoubtedly huge for all of the Congo Basin countries. As such, the corridor approach has also been adopted by regional entities, such as the Economic Community of Central African States (ECCAS) and CEMAC (Economic and Monetary Community of Central Africa) to foster synergies and economies of scale among their member states (CEMAC 2009).

Improve Land Tenure Schemes

Current land tenure schemes are not conducive for SFM in the Congo Basin countries. Outside of commercial logging concessions, forests are considered as "free access" areas under state ownership and not tagged with property right. Moreover, tenure laws in most Basin countries directly link forest clearing (*mise en valeur*) with land property recognition. This results in an incentive to expand agriculture into forested areas and promotes the unsustainable exploitation of woodfuel, as the resource itself is drastically underpriced.

Governments should prioritize improving land tenure and land use schemes in order to stimulate management of natural resources and reduce pressures on primary forests. Congo Basin countries must strengthen their rural land governance and tenure recognition framework. Effective systems of land use and access rights and, in general, property rights are essential: they ameliorate the management of natural resources and stimulate sustainable agriculture. Improving them is a priority for providing farmers, especially women, with the incentives needed to make long-term investments in agricultural transformation. With respect to woodfuel production, community-based forest management approaches can successfully expand the supply-and-relief natural forests from unsustainable withdrawals. However, communities will invest in sustainable forest practices or tree plantations/agroforestry systems only if they are given enough visibility on land/tree tenure issues. In general terms, clarification of land/tree rights is considered a key prerequisite to any actions that support SFM practices and should be given a high priority.

Clarifying land tenure can also help create an enabling environment for responsible private investment. In addition to allowing farmers to invest in their land, the clarification of land rights on their whole territory would also permit Basin countries to become more proactive and engage in fairer negotiations with potential large investors. Weak land governance poses a risk that investors will acquire land almost for free and neglect local rights or environmental issues, with potentially far-reaching negative consequences[5] (Deininger et al. 2011). In addition, a strong correlation has been evidenced between large land applications and the weakness of rural land tenure recognition in the target countries, which clearly suggests that Basin countries are at risk (Deininger and Byerlee 2011). Accordingly, governments should establish stronger policies on future large land investments, including requiring land applications be oriented toward abandoned plantations and suitable nonforested land, and requiring environmental impact assessments (EIAs).

Strengthen Institutions and Law Enforcement

Strengthening institutions in the Congo Basin is a priority. Without strong institutions to enforce rules and build alliances within a complex political economy, neither land use planning nor tenure reform, as described above, will yield real change. A particular focus should be given to forestry administrations that are generally weak in all Basin countries, where they are often understaffed or have older staffs. The expectations in terms of planning, monitoring, and controlling forest resources as well as enforcing law are high but largely decoupled with the existing capacities. To effectively protect and manage forests, a legal framework needs to be put in place and then fully enforced by properly staffed and equipped institutions. This is of particular importance in the fight against illegal activities but also in the formalization process that should be engaged with the artisanal logging sector as well as with the informal woodfuel/charcoal value chain (see below sections related to specific sectors).

Staff in forestry administration is generally highly concentrated in headquarters and central entities, with very few people at the decentralized level. The staff has usually not been trained in new techniques, technologies, and dimensions of forest management. Beyond human resources, forestry administrations are also poorly equipped, particularly in decentralized offices. Accordingly, priority should be given to the following aspects:

- Rejuvenating forestry staff: Staffing strategies (recruitment and capacity building) for the forestry administration should be redefined based on new needs in terms of knowledge and skills.
- Fostering technology transfer: Administration usually relies on inadequate equipment and buildings. New technologies (log tracking system, geographic information system [GIS], and so on) should be transferred to forestry administrations so that they can perform more efficiently their core tasks in terms of planning, monitoring, and control.

To succeed, REDD+ needs to build on strong institutions, notably in terms of law enforcement and monitoring. REDD+ is likely to fail if the basic governance issues are not properly addressed. Congo Basin countries have comparatively weak institutions and governance frameworks. They are best advised to develop REDD+ strategies that account for the particular circumstances and challenges of the region (see box 3.3). Furthermore, to get ready for phase 3 financing for REDD+ institutions in the Congo Basin, countries will have to be able to set up credible monitoring systems so the international community can track progress made in specific countries. The dire lack of data will have to be overcome, and appropriate technologies that take into account the specific challenges—difficult access to forested areas, understaffed decentralized institutions, and so on—must be deployed.

Fostering multisectoral coordination

By looking at opportunities to mitigate GHG emissions at the landscape level, REDD+ could emerge as a development planning approach to coordinate forests and other land uses. The implementation of REDD+ strategy and, more broadly, national land use strategies requires strong coordination among the different line ministries, local institutions, and other stakeholders. This coordination is usually weak, and new schemes will have to be defined with strong high-level support to ensure coordination between the various sectors involved in REDD+.

Building strategic partnerships

Corporate compliance with national laws and regulations must be monitored by regulatory agencies, but such oversight usually proves challenging in central African countries, where lack of capacity, inaccessibility of some sites, governance issues, and security risks can make regulation difficult at best. Where possible, strategic partnerships can improve monitoring activities at the local level. Communities can be trained and engaged in helping regulators monitor activities on the ground; nongovernmental organizations can also provide additional

---

Box 3.3 Fragile States and REDD+ Challenges

The originality of the REDD+ proposal is its incentive-based mechanism designed to reward the governments of developing countries for their performance in reducing deforestation as measured against a baseline. This mechanism is founded on the hypothesis that developing countries "pay" an opportunity cost to conserve their forests and would prefer other choices and to convert their woodlands into other uses. The basic idea is, therefore, to pay rents to these countries to compensate them for the anticipated foregone revenues. The reference to the theory of incentives (in its principal–agent version) is implicit but clear. In this REDD+-related framework, the government is taken as any economic agent that behaves rationally—that is, making decisions after comparing the relative prices associated with several alternatives, deciding on action, and then implementing effective measures to tackle deforestation and shift the nationwide development path.

Karsenty and Ongolo (2012) argue that such an approach ignores the political economy of the state, especially when dealing with "fragile" or even "failing" states that face severe but chronic institutional crises, which are often ruled by "governments with private agendas" fueling corruption. Two assumptions underlying the REDD+ proposal are particularly critical:

- The idea that the government of such a state is in a position to make a decision to shift its development pathway on the basis of a cost–benefit analysis that anticipates financial rewards.
- The idea that, once such a decision has been made, the "fragile" state is capable, from the financial rewards, to implement and enforce the appropriate policies and measures that could translate into deforestation reduction.

---

monitoring via field projects. In the context of the agricultural sector and its potential impacts on forested areas, the Comprehensive Africa Agriculture Development Program process offers an excellent and timely opportunity to thoroughly analyze agricultural potential; develop or update national and regional agricultural investment plans aimed at increasing agricultural productivity on a sustainable basis; and strengthen agricultural policies. For the forest sector, the REDD+ readiness and Forest Law Enforcement, Governance, and Trade (FLEGT) processes afford platforms for coordination and building strategy.

Agriculture: Increase Productivity and Prioritize Nonforested Lands

Subsistence agriculture is considered one of the major drivers of deforestation in the Congo Basin now. Agroindustry plantations may become another engine of deforestation in the coming years. More recognition is given to taking a sustainable "landscape" approach when considering REDD+ strategies, with a particular focus on the need for appropriate incentives to farmers and communities living on the forest frontier. Below are a series of recommendations that could help reconcile agricultural production increase and preservation of primary forests. The recommendations come in addition to those indicated in the above sections, which are crosscutting and apply to any land use sector.

- **Prioritize agricultural expansion on nonforested areas, based on a participatory land use planning exercise.** There is an estimated 40 million hectares of suitable noncropped, nonforested, and unprotected land in the Congo Basin. This amount corresponds to more than 1.6 times the current area under cultivation. Using these available areas, along with the increase in land productivity, could dramatically transform agriculture in the Basin without taking a toll on forests. Decision makers must prioritize the expansion of agriculture on nonforested lands.

- **Promote "climate-smart" agriculture that increases productivity while reducing vulnerability.** In Basin countries, "climate-smart" agriculture would mainly take the form of conservation agriculture or minimal soil disturbance—for example, refraining from tillage and direct seeding, maintaining a mulch of carbon-rich organic matter that protects and feeds the soil, using rotations and associations of such crops as trees that would include nitrogen-fixing legumes and agroforestry (intensive use of trees and shrubs in agricultural production). The latter has been developed at the pilot level around major urban centers, such as Kinshasa, to respond simultaneously to rising needs in food and energy (see box 3.4).

Box 3.4 Feeding Cities: Mixing Charcoal and Agroforestry in Kinshasa

Kinshasa, a megacity of 8 to 10 million inhabitants, is located in a forest–savanna mosaic environment on the Batéké Plateau in the Democratic Republic of Congo. The city's wood energy supply of about 5 million cubic meters per year is mostly informally harvested from degraded forest galleries within a radius of 200 kilometers of the city. With gallery forests most affected by degradation from wood harvesting, even forests beyond this radius are experiencing gradual degradation, while the periurban area within a radius of 50 kilometers of Kinshasa has suffered total deforestation.

However, plantations around the megacity help provide wood energy on a more sustainable basis along with food. About 8,000 hectares of plantations were established in the late 1980s and early 1990s in Mampu, in the degraded savanna grasslands 140 kilometers from Kinshasa, to meet the city's charcoal needs. Today the plantation is managed in 25-hectare plots by 300 households in a crop rotation that takes advantage of acacia trees' nitrogen-fixing properties and the residue from charcoal production to increase crop yields. Another scheme, run by a Congolese private company called Novacel, intercrops cassava with acacia trees in order to generate food and sustainable charcoal, as well as carbon credits. To date, about 1,500 hectares out of a projected 4,200 have been planted. The trees are not yet mature enough to produce charcoal, but cassava has been harvested, processed, and sold for several years. The company has also received some initial carbon payments. The project has been producing about 45 tons of cassava tubers per week and generates 30 full-time jobs, plus 200 seasonal jobs. Novacel reinvests part of its revenue from carbon credits into local social services, including the maintenance of an elementary school and a health clinic.

- **Empower smallholder farmers.** With about half the population active in agriculture in most countries of the Congo Basin, there is a need to foster sustained agricultural growth based on smallholder involvement. Experience in other tropical regions shows this is possible (Deininger et al. 2011). Thailand, for example, considerably expanded its rice production area and became a major exporter of other commodities (sugar, cassava, and maize) by engaging its smallholders through a massive land titling program and government support to research, extension, credit, producer organizations, and rail and road infrastructure development.

  New incentives schemes should be set up for smallholder farmers, especially when the adoption of new practices implies a loss of income in the first years, possibly through payments for environmental services. At the country level, access to credit or provisions in-kind (including access to land, markets, or production inputs) could be established to stimulate the adoption of sustainable agricultural practices. At a broader level, market-based incentives could be created, through certification schemes, to support large and small producers in large agroindustry (that is, oil palm, rubber, and so on), which adhere to sustainable practices. Complementary to positive incentives, it is also crucial to make sure that measures with potential adverse impacts on forests be removed. Such negative incentives can include regulatory provisions that link property rights on land with forest clearing (see Improve Land Tenure Schemes above) or credit schemes offered by commercial banks to support activities that require deforestation. Removing such perverse incentives has proven to be particularly efficient in terms of curbing deforestation. In Brazil, the veto from the Banco do Brasil on agricultural credit for farmers who wanted to clear areas of the Amazon forest immediately reduced the pressure on forests.

- **Reinvigorate research and development (R&D) toward sustainable productivity increases.** R&D capacities in the Congo Basin, with the exception of Cameroon, have been dismantled over the past decades. National research centers are dysfunctional and unable to take over the challenge to transform the agriculture sector. Research has largely neglected the most common staple crops, such as yams, plantains, and cassava, which are of particular importance in the Basin; they are usually referred to as "neglected crops." The potential to increase productivity of these crops and to improve both their resistance to disease and their tolerance to climatic events has been untapped so far.

  Agricultural R&D in the Congo Basin needs to be stimulated through partnerships with international research centers (for example, among members of the Consultative Group on International Agricultural Research [CGIAR]) whose objective is to progressively strengthen national capacities. In addition to R&D, extension services will also need to be revitalized to mainstream new agricultural practices in the rural areas. Mechanization should be supported to boost smallholder labor productivity; this includes improving current labor-intensive postharvest operations that are often staffed by women.

- **Promote a sustainable large-scale commercial industry through improved regulations, particularly on procedures for land allocation and environmental management.** Large agribusiness operations—especially rubber, palm oil, and sugarcane plantations—could sustain economic growth and generate considerable employment for rural populations. Moreover, large companies' ability to readily overcome the market imperfections prevailing in Basin countries—especially in regard to access to finance, technology, inputs, processing, and markets—makes them a potentially important actor in a sustainable agricultural development strategy. Large agribusiness operations can also play a positive role in reducing deforestation and forest degradation by employing relatively large populations who would, therefore, forgo their traditional slash-and-burn practices. In addition, they also have a legal obligation in most Congo Basin countries to provide social infrastructures (schools, hospitals, and so on).

  However, in order to achieve this end, adequate policy, regulatory, and institutional capacities need to be in place to mitigate environmental and social risks associated with large private land development investments. Feeble governance presents a risk that investors will acquire land almost for free, interfere with local rights, and neglect their social and environmental responsibilities. Although Basin forests have been largely exempted from significant "land grabbing" attempts up to now, weak governance could imperil the Congo Basin in the future. In fact, the risks are considered to be very high that large-scale operations often get access to natural resources with insufficient attention paid to social and environmental externalities and no attempt to maximize the potential impact of private investment on poverty reduction (Deininger et al. 2011).

  Governments should establish stronger policies on future large land investments, including requiring land applications to be oriented toward abandoned plantations and suitable nonforested land. Initiatives to put palm oil production on a more sustainable footing, such as the Roundtable on Sustainable Palm Oil founded in 2004, may help mitigate some of these environmental issues by setting standards that prevent further loss to primary forests or high conservation value areas, and reduce impacts on biodiversity.

- **Foster win–win partnerships between large-scale operators and smallholders.** Such partnerships could make the current dualistic profile of agriculture (small and large scales) in the Congo Basin an engine for transforming the agricultural sector. While this situation has not yet materialized in the Basin, there are many examples of it elsewhere, and meaningful partnerships between smallholders and large-scale operators have yielded successful results and promoted a well-balanced development of agriculture. Innovative and Congo Basin–specific outsourcing schemes could be piloted and duplicated.

Box 3.5 Partnerships between Large-scale Operators and Smallholders: Examples

In Indonesia, which is now the world's largest palm oil producer, smallholders account for about a third of the country's production. Because of processing requirements and the rapid deterioration of fresh fruit, and poor access to capital and planting material, most small oil palm producers are in formal partnerships with oil palm companies through nucleus/out-grower schemes. Average income from oil palm cultivation is higher than from subsistence farming or competing cash crops, and it is estimated that oil palm expansion in Indonesia significantly helped reduce rural poverty.

Rubber was originally grown on large plantations in humid forest areas of Southeast Asia but then, because of rising labor and land costs, increasingly became a smallholder production. Farms of 2-3 ha now make up 80 percent of world production. That was made possible by the development of improved hevea clones and techniques suited to production and processing at the smallholder level. Smallholders in Indonesia produce rubber in improved agroforestry systems that maintain carbon stocks and species richness. While returns from such systems are lower than those of monocultures, reduced risk and lower initial capital costs more than compensate, and efforts are underway to certify rubber from these systems to obtain a price premium.

Wood-Based Energy: Organize the Informal Value Chain

Heavy reliance on wood extraction for domestic woodfuel or charcoal production puts huge pressure on natural forests in the Congo Basin, particularly in densely populated areas. In low-population rural areas, woodfuel is often a sustainable resource; urban areas have vastly overexploited it. There is an urgent need to organize the charcoal value chain with a sustainable wood supply. A REDD+ mechanism could present an opportunity to modernize this segment of the energy sector. Following is a series of relevant recommendations; these recommendations come in addition to those previously indicated that are cross-cutting and apply to any land use sector.

- **Place the woodfuel energy sector higher on the political agenda**. The importance of this source of energy in Africa is undisputed; however, until now, very little attention has been paid to this sector in the policy dialogue. Therefore, it is poorly featured in official energy policies and strategies. Reasons for this situation include the following: (i) the wood-based energy sector is perceived as "old-fashioned" and "backward," and policy makers are instead interested in more modern and supposedly cleaner sources of energy; (ii) the wood-based energy sector, usually associated with forest degradation and deforestation, is seen as a harmful sector that needs to be eradicated; (iii) the sector is poorly documented and does not have the benefit of any reliable statistical data (which tends to minimize its role in terms of contribution to economic growth, employment); (iv) the sector's mostly informal governance often motivates rent-seeking behaviors and conflicting interests, which may hamper reforms.

Policy makers' perception of wood energy as "traditional" and "old-fashioned" needs to change. Lessons could be drawn from Europe and North Africa, which place wood energy among the most modern energy sources (for heating, electrical power, and sometimes cooking).

- **Formalize the woodfuel/charcoal value chain**, which would break the oligopolistic structure of the sector and open up to a more transparent marketplace. The economic value of the resources will thus be better reflected in the pricing structure, and appropriate incentives could be established. Such formalization should be supported by the revision and modernization of the regulatory framework. To do so, priority should be given to the following aspects: (i) data collection and research to fully understand the nature and impact of woodfuel collection on forests; (ii) comprehension of the "political economy" of the informal woodfuel/charcoal value chain; and (iii) development of a cross-sectoral woodfuel strategy and, accordingly, adjustment of the legal and regulatory framework in a participatory/consensual manner.

- **Diversify the supply side of the value chain.** The charcoal value chain in the Congo Basin currently relies on natural forests exclusively. Although natural forests are expected to continue supplying much of the raw material for charcoal production, they will be unable to meet demand in a sustainable manner because it is expected to increase substantially. There is a need to ensure that the whole charcoal value chain properly integrates the sustainability dimension of the wood supply. In order to do this, policy makers should consider the two following options: (i) maximizing the potential of sustainable harvests from natural forests, with a special consideration to timber waste management; and (ii) increasing sustainable wood supply through tree plantations and agroforestry.

- **Foster community involvement through devolution of rights and capacity building.** Community-based woodfuel production schemes in Niger, Senegal, Rwanda and Madagascar have shown promising results when long-term rights to forest land and devolution of management have motivated communities to participate in woodfuel production. As indicated in above-section on "Improve land tenure schemes", priority should be given to the reform of the land tenure scheme in order to enhance access and rights security of local communities over lands. Pilots have been launched in the Congo Basin (Batéké plantations) and could be replicated.

- **Respond to growing urban needs in terms of both food and energy.** Deforestation and forest degradation chiefly occur around the urban centers in the Basin countries, mostly due to anarchical agricultural expansion in response to rising demand for food and energy. Periurban agriculture deserves special attention through an integrated approach that would address the various driving forces of forest degradation. If well organized, it could not only secure the food and energy provision for a growing urban population in most Congo Basin countries, but it could also provide sustainable solutions to unemployment and waste management.

Transport: Better Plan to Minimize Adverse Impacts

With decaying roads, rails, and ports, the Congo Basin has one of the world's most severe infrastructure deficits, which has stymied development efforts and led to fragmented economies. Countries in the region have placed new transportation infrastructure high on the political agenda, with the objective to increase market access, decrease the price of imports, and raise the competitiveness of local products for export. Yet transportation infrastructure can lead to significant deforestation if poorly planned. Although direct impacts on forests from transportation development are not extensive, the indirect and induced impacts can be severe and widespread. A system that provides incentives for REDD+ can support a development plan with a more integrated approach for transport infrastructure. Following is a series of recommendations; these recommendations come in addition to those previously indicated that are crosscutting and apply to any land use sector.

- **Improve transportation planning at local, national, and regional levels**. Developing transportation infrastructure while mitigating deforestation requires a thorough reflection on the development model at all levels. This is necessary at the local level because areas that are directly served by improved transportation facilities will become more competitive for various economic activities (such as agricultural expansion, including palm oil plantations). This is essential at national and regional levels because the corridor approach shows that improving transportation services (for example, freight management in harbors) or infrastructure (facilitating river or rail transportation) may have a globally wider macroeconomic impact.

  Local participation in transportation planning will help ensure that economic opportunities are maximized. Mitigation measures at the local level may include clarifying land tenure or integrating the transportation project into a broader local development plan. Such plans may include the protection of forest banks along roads, rivers, or railways to avoid unplanned deforestation. Planning at the national and regional levels, through a corridor approach, could help identify adequate mitigation measures, such as zoning reforms (establishing permanent forest areas), law enforcement (ensuring the respect of zoning decisions), land tenure clarification, and controlling the expansion of agriculture.

- **Foster a multimodal transport network**. Although much focus is given to roads, other modal systems can support economic growth in the Congo Basin. For instance, with more than 12,000 kilometers of navigable network, the

Basin could benefit from a potentially highly competitive waterway system; however, river transportation fails to live up to the role it could play in overall economic development in the Congo Basin. As a result, despite the vast potential, the waterway system remains a marginal transport mode there. The same situation applies to the railway system, to a lesser extent (particularly for passenger transportation). While countries plan for transport development, it is important that they consider alternative modes and consider the pros and cons of the different alternatives, not only in terms of economic returns but also in terms of environmental impacts.

- **Properly assess ex ante impacts of transport investments.** Transport development (whether new infrastructure or rehabilitation of the existing ones) will reshape the economic profile of the impacted areas and will consequently increase pressure on forest resources, if any. Currently, most of the environmental impact studies or safeguard reviews fail to fully capture the long-term indirect effects on deforestation. Therefore, there is a need to develop a new set of instruments that would help determine the impact of increasing economic competitiveness in the areas served by new transportation infrastructure. A robust ex ante assessment of the potential indirect and induced effects of transport development should be an integral part of the design phase of the infrastructure investments and could help design the mitigation measures. To do so, a robust economic modeling exercise (economic prospective analysis) should be undertaken as part of any infrastructure investment preparation. This would ensure that transportation investments are designed consistently with a low-impact economic development.

Logging: Expand SFM to Informal Sector

Logging activities result in forest degradation rather than deforestation in the Congo Basin. Although progress has been made in industrial logging concessions, there is still room for improvement, and efforts should be pursued. However, it is clear that the major threat from logging activities now comes from the informal sector, which is not ruled by any governance framework and which tends to adversely impact forest resources. REDD+ provides new momentum to revisit the status of SFM in the Congo Basin with differentiated approaches inside and outside commercial forest concessions. Following is a series of recommendations; these recommendations come in addition to those previously indicated that are crosscutting and apply to any land use sector.

- **Pursue progress on SFM in industrial logging concessions.** The Congo Basin region is one of the most advanced in terms of areas with an approved (or under preparation) management plan (MP). However, studies indicate that SFM principles have yet to fully materialize at the level of industrial logging concessions. Particular attention should be paid to the following aspects: (i) ensure adequate implementation of MPs at the concession level; (ii) adjust SFM standards and criteria; and (iii) move away from single-use, timber-oriented management models. Forest certification schemes could also be encouraged[6] as in many places—communities' perceptions of improved social benefits from certified concessions are higher than those of noncertified ones.

- **Further operationalize the "community forest" concept.** Community forestry has been embraced by most of the Basin countries and is now reflected in their legal frameworks; however, challenges remain in terms of its operationalization. Community forest status usually does not carry permanent property rights and is de facto similar to a concession, simply smaller and under a different regulatory framework. Tenure rights issues and various other hurdles significantly limit opportunities for effective and sustainable community management of forest resources.

- **Support the FLEGT Process.** The Forest Law Enforcement, Governance, and Trade, led by the European Union, is the most comprehensive initiative to support tropical timber-producing countries' efforts to improve governance in their forestry. The process is already quite advanced in all Basin countries (except Equatorial Guinea), and any forest governance-related activities in a specific country should strengthen and be aligned with the FLEGT Voluntary Partnership Agreement (VPA) signed by this country.

- **Formalize the informal timber sector, which has been long overlooked.** The expansion of this subsector has recently been driven by booming domestic markets and is by far uncontrolled and unregulated, creating major pressure on natural forests. To ensure a sustainable timber supply of domestic markets, the myriad small and medium forest enterprises would need to be supported by appropriate regulations.

  In order to create an appropriate regulatory framework for domestic timber production, the Basin countries' governments should carefully consider the following: (i) understanding the "political economy" of the informal timber value chain, and then accordingly (ii) adapting the legal and regulatory framework to move the informal sector into formality. Trends of new markets (domestic and regional) should also be better understood.[7]

  It is essential for the governments, as they prepare new frameworks on domestic timber production and trade to engage in an open and transparent dialogue with all key stakeholders and particularly the local people who benefit from the informal activities. A multistakeholder dialogue will be critical for identifying solutions to the difficult trade-offs between sustaining rural livelihoods dependent on the informal domestic markets and enforcing production standards and trade restrictions as required by the principles of timber legality.

- **Modernize the processing capacities.** Having a higher performing and modern timber-processing industry has always been a top priority for the Congo Basin governments. To set up an efficient timber value chain in the Congo

Basin, updating the processing sector is crucial. The following dimensions should be taken into account:

- Adjusting the processing capacities to the forest resources (taking into consideration both export-oriented and domestic markets). An alternative timber supply can also be considered through plantations and an "on-farm trees" system.
- Promoting more efficient processing techniques. Basin countries' governments should first have a comprehensive assessment of the barriers that need to be lifted in order to promote more efficient in-country processing (beyond primary processing).
- Diversifying valorized species through access to new consuming markets (beyond European and Asian markets that tend to be selective) and through more advanced transformation processes (such as the incorporation into plywood production and other secondary processing).

Mining: Set "High-Standard" Goals for Environmental Management

To date, mining activities have had limited impacts on Basin forests because the majority of the mining operations have occurred in nonforested areas; however, expanded mining activities will increasingly impact the forest. A REDD+ mechanism could accompany the development of mining activities in the Congo Basin and minimize adverse impacts on natural forests. Following is a series of recommendations; these recommendations come in addition to those indicated in the above sections that are crosscutting and apply to any land use sector.

- **Properly assess and monitor impacts of mining activities at all stages (from exploration to mine closure).** Proper EIAs and social impact assessments (SIAs) have to be prepared for all of the different stages and for MPs to mitigate the associated risks. In many countries, the EIAs/SIAs are now required by law.[8] However, very often these assessments do not meet the minimum quality standard. Beyond weaknesses in the assessments themselves, the associated mitigation plans are often poorly designed and difficult to implement and monitor.

- **Learn from international best practices and foster risk mitigation.** If mining is to result in minimal negative impacts on the forests of the Congo Basin, companies will need to follow best international practices and standards designed to meet the mitigation hierarchy (avoid, minimize, restore, and compensate). International standards for responsible mining have been developed by varying (and sometimes competing) organizations, including the International Council on Mining and Metals, the Responsible Jewelry Council, the International Finance Corporation, and the Initiative for Responsible Mining Assurance. Although these initiatives address large-scale mining, a corollary also exists for the small-scale mining sector—the Alliance for Responsible Mining (ARM). Indeed, the ARM has developed a certification system for small-scale mining cooperatives that includes consideration of both environmental and social concerns. In addition, some oil and gas companies (for example, Shell Oil) have already developed extractive projects that seek to minimize the impact of extraction on forests in Basin countries. Lessons can be learned from these innovative approaches as the governments adjust their national regulations on mining activities and their environmental monitoring and management.

- **Upgrade the small-scale mining sector**. In many cases, the impacts of artisanal mining activities—though scattered and more difficult to assess and monitor—are expected to be significant, specifically through the cumulative effects over time. In some countries, hot spots of deforestation are clearly linked to small-scale mining activities. Efforts should be focused on organizing small-scale miners and adjusting the regulatory frameworks so that they can better respond to specific needs. Governments should facilitate the use of environmentally friendly technologies, such as retorts and other mercury-capturing devices. ARM's principles and standards (especially Standard Zero) could form the basis for regulations and incentives to improve the environmental performance of this sector.

- **Promote innovative mechanisms to offset negative impacts of mining operations**. In addition to minimizing negative impacts from the extractive industries, creating net positive gain from mining development is getting more attention. Among the options under consideration is the creation of biodiversity offsets. Conservation groups have advocated for biodiversity

---

Box 3.6 In Search of Green Gold

Both artisanal miners (who operate with little mechanized aid) and small-scale miners (who use more organized and more productive methods but produce less than a certain tonnage of minerals per year) have responded to international demand for minerals by increasing activity in the Congo Basin in recent years. Some of the environmental concerns associated with artisanal and small-scale mining stem from practices that can include primary forest clearance, dam construction, the digging of deep pits without backfilling, and resulting impacts on water levels and watercourses. Forest degradation is also associated with the arrival of large numbers of migrant diggers spread out over a large area of forest. In Gabon, for example, artisanal miners suffer from a fragile legal status that gives them little incentive to pursue environmentally responsible mining (WWF 2012).

Strategies to respond to these issues include the setting up of socially responsive and environmentally sustainable supply chains, and measures to professionalize and formalize artisanal and small-scale mining activities so that risks are managed and minimum standards introduced. These initiatives are partially inspired by the success of a third-party certification scheme called "Green Gold – Oro Verde," born in 1999 in Colombia to stop the social and environmental devastation caused by poor mining practices in the lush Chocó Bioregion, and to supply select jewelers with traceable, sustainable metals.

---

offsets for extractive projects for at least a decade.[9] But the concept lacked practical implementation, and companies complained of a dearth of guidelines for biodiversity offsets. Financial instruments, such as financial guarantee, could also mitigate adverse impacts, particularly to ensure mine reclamation and restoration at the closure of the mining site.

Notes

1. FCCC/CP/2010/7/Add.1, Decision 1/CP/16, para 73.

2. Beyond bilateral and multilateral financing, domestic financing is in some instances also significant, particularly in emerging and middle-income economies. Brazil, for example, reports an historical annual average of US$500 million for monitoring and inventory work, law enforcement, tenure reform, as well as national and local plans to reduce deforestation. Mexico, Costa Rica, and Indonesia also use domestic finance to support REDD+ activities.

3. In nonforest sectors, reference levels are typically given as the net emissions in a particular year. Improvement over time, measured by lowering the net emissions in subsequent years, is how performance is measured. For the Kyoto Protocol, the "base year" was set at 1990.

4. The Global Partnership on Forest Landscape Restoration estimates that more than 400 million hectares of degraded land in Sub-Saharan Africa offer opportunities for restoring or enhancing the functionality of "mosaic" landscapes that mix forest, agriculture, and other land uses. Accessible at http://www.ideastransformlandscapes.org/.

5. For example, on-the-ground verification of recent land acquisitions in the Democratic Republic of Congo has evidenced irregularities in land allocation processes: although all concessions of at least 1,000 hectares must be approved by the Minister of Land Affairs, data collection in the Katanga and Kinshasa provinces suggested that governors have in some cases awarded multiple concessions of up to 1,000 hectares each to individual investors who circumvent the required approval procedure (Deininger et al. 2011).

6. Although adopting certification schemes is a voluntary process, incentives could be arranged in the Congo Basin governments to encourage private operators to choose to certify their concessions.

7. Analysis of these "new markets/flows" should be preferably conducted at the regional level, as there are clear signals that the timber flows tend to be transnational.

8. The exploration phase is, however, generally not covered by environmental impact assessments (EIAs) in Basin countries even though significant impacts can occur at this stage.

9. See, for example, Conservation International 2003.

References

Angelsen, A., D. Boucher, S. Brown, V. Merckx, C. Streck, and D. Zarin. 2011a. *Guidelines for REDD+ Reference Levels: Principles and Recommendations*. Washington, DC: The Meridian Institute.

————. 2011b. *Modalities for REDD+ Reference Levels: Technical and Procedural Issues.* Washington, DC: The Meridian Institute.

CEMAC (Economic and Monetary Community of Central Africa). 2009. *CEMAC 2025: Towards an integrated emerging regional economy: Regional Economic Program 2010–2015 (Vers une économie régionale intégrée et émergente Programme Economique Régional 2010–2015).* Volume 2. CEMAC, Bangui, Central African Republic.

Conservation International. 2003. *Opportunities for Benefiting Biodiversity Conservation. The Energy and Biodiversity Initiative.* Washington, DC: Conservation International. http://www.theebi.org/pdfs/opportunities.pdf.

Deininger, K., and D. Byerlee. 2011. "The Rise of Large Farms in Land Abundant Countries: Do They Have a Future?" Research Working Paper 5588, World Bank, Washington, DC.

Deininger, K., D. Byerlee, J. Lindsay, A. Norton, H. Selod, and M. Stickler. 2011. *Rising Global Interest in Farmland—Can It Yield Sustainable and Equitable Benefits?* Washington, DC: World Bank.

Gledhill, R., C. Streck, S. Maginnis, and S. Brown. 2011. "Funding for Forests: UK Government Support for REDD+." Joint report, PricewaterhouseCoopers LLP/Climate Focus/International Union for Conservation of Nature (IUCN)/Winrock International.

Karsenty, A., and S. Ongolo. 2012. "Can 'Fragile States' Decide to Reduce Their Deforestation? The Inappropriate Use of the Theory of Incentives with Respect to the REDD Mechanism." *Forest Policy and Economics* 18 (May): 38–45.

Martinet, A., C. Megevand, and C. Streck. 2009. *REDD+ Reference Levels and Drivers of Deforestation in Congo Basin Countries.* Washington, DC: World Bank and Forestry Commission of Central Africa.

Conclusions and Outlook

The countries of the Congo Basin face the dual challenge of developing local economies and reducing poverty while limiting the negative impact of growth on the region's natural capital and, particularly, forests.

Development needs are great. Despite abundant natural assets, the percentage of the population beneath the national poverty line hovers between one-third and two-thirds of the population in different countries of the Basin, access to food is majorly inadequate, and undernourishment is highly prevalent. Transportation infrastructure is among the most deteriorated in the world creating de facto a juxtaposition of landlocked economies within the region, which considerably accrue farmers' vulnerability to poor harvests. Looking ahead, the Congo Basin population is expected to double between 2000 and 2030, leading to a total of 170 million people by 2030—people in need of food, energy, shelter, and employment.

Natural assets have so far been largely preserved; deforestation rates in the Congo Basin are among the lowest in the tropical rainforest belt and are significantly below the deforestation rates experienced by most other African regions. The canopy has benefitted to some extent from "passive protection" provided by political instability and the lack of transportation infrastructure.

However, this situation may change. Although subsistence activities such as small-scale agriculture and woodfuel collection are currently the main causes of deforestation and degradation in the Congo Basin, new threats are expected to emerge and aggravate the pressures on natural forests. In local and regional development, population increases and global demand for commodities are expected to jointly drive accelerated deforestation and forest degradation, if "business as usual" models are to be applied.

Congo Basin countries are now at a crossroad—they are not yet locked into a development path that will necessarily come at a high cost to forests. They can define a new path toward "forest-friendly" growth. The question is how to

accompany economic change with smart measures and policy choices so that
Congo Basin countries sustain and benefit from their extraordinary natural assets
over the long term. In other words, how to "leapfrog" the traditional dip in forest
cover usually observed in the forest transition curve.

New environmental finance mechanisms can help Congo Basin countries to
transition toward a forest-friendly development path. Environmental finance
includes climate funding for adaptation and mitigation efforts in general, and
reducing emissions from deforestation and forest degradation plus (REDD+) in
particular, but also financing for biodiversity, wetlands, or soil restoration. When
accessing these new resources, countries may consider a number of issues in order
to prioritize activities and effectively allocate these new funds. It is therefore up
to national governments to define how these various mechanisms fit into their
own development, how to best use such resources, and whether and how to meet
the relevant criteria of funds or mechanisms, and to assess the benefits and risks
associated with particular funds, including the costs of putting into place relevant
information and institutional conditions.

REDD+ provides an important opportunity for Congo Basin countries to
develop strategies that work toward sustainable development while protecting
the natural and cultural heritage of the region. This new, dedicated focus on forest protection within international climate agreements, in combination with the
availability of significant new financial resources, moves sustainable forest management up in the political agenda and has facilitated in many countries a dialogue among forest agencies and those ministries and entities that regulate
broader industrial and agricultural development.

However, conditions and scale of eventual REDD+ financing remain
uncertain. In particular, it remains unclear how "results-based" financing will be
measured, what the criteria for payments might be, and how much funding will
be made available under such a scheme. To date, these issues have not been
clarified by international negotiations, neither the rules that will guide the
establishment of national reference levels nor the reference emissions levels that
would allow fully measured results-based finance. In the near to medium term,
there will likely be a multiplicity of donors and fragmentation of REDD+
finance—including a fragmented REDD+ market. In this complex landscape of
finance, it is important governments prioritize activities, partnerships, and
processes. Engagement with each donor, and their specific requirement, or each
process related to multilateral funding or emerging carbon markets, requires
significant resources.

Nevertheless, there are resources available now that countries can use for "no
regrets" measures. Such measures, while differing from country to country,
should seek to create the enabling conditions for the implementation of an
inclusive, green growth. The study *Deforestation Trends in the Congo Basin:
Reconciling Economic Growth and Forest Protection* highlights a number of no
regrets actions that countries can take now to grow along a sustainable development path. The following recommendations have emerged from the technical
discussions among experts, including experts from the Congo Basin countries at

the regional level. These recommendations are intended as general guidelines to spur more detailed discussions at the country level.

- Participatory land use planning could help clarify trade-offs among different sectors, encourage the development of growth poles and corridors, and direct destructive activities away from forests of great ecological value.
- Institutions need to be strengthened to adequately perform their regalian responsibilities in terms of planning, monitoring, and law enforcement. Multisectoral coordination should be promoted to foster an integrated vision for national development.
- Unlocking the potential of the Congo Basin for agriculture will not necessarily take a toll on forests: the Congo Basin could almost double its cultivated area without converting any forested areas. Policy makers should seek to primarily target agricultural activities toward degraded and nonforested land.
- In the energy sector, putting the woodfuel supply chain on a more sustainable and formal basis should stand as a priority. More attention should be paid to respond to growing urban needs for both food and energy through intensified multiuse systems (agroforestry).
- Better planning at the regional and national levels could help contain the adverse effects of transportation development, through a multimodal and more spatially efficient network.
- Expanding sustainable forest management principles to the booming and unregulated informal logging sector would help preserve forest biomass and carbon stocks.
- Setting "high-standard" goals for environmental management of the mining sector could help mitigate adverse effects as the sector develops in the Congo Basin.

Congo Basin countries will need to find ways to strategically make use of the multiple sources of funds. The multiple sources of financial assistance available to developing countries—REDD+ and climate change mitigation, climate change adaptation, carbon finance, biodiversity, food security, or general development finance—can be frustratingly complex, each with its own set of requirements and criteria. The effective use of funds will require close coordination among national and international actors, clear planning, and the development of an integrated and supportive set of policies. A national low-carbon, climate-resilient strategy—which includes the protection of forest resources and addresses the key drivers of deforestation in a holistic manner—can provide this clear pathway to effective development in the region.

GLOBIOM Model—Formal Description

Objective function

$$
\begin{aligned}
\text{Max } WELF_t = & \sum_{r,y}\left[\int \phi_{r,t,y}^{\text{demd}}\left(D_{r,t,y}\right)d\left(\cdot\right)\right] - \sum_{r}\left[\int \phi_{r,t}^{\text{splw}}\left(W_{r,t}\right)d\left(\cdot\right)\right] \\
& -\sum_{r,l,\bar{l}}\left[\int \phi_{r,l,\bar{l},t}^{\text{lucc}}\left(\sum_{c,o,p,q}Q_{r,t,c,o,l,\bar{l}}\right)d(\cdot)\right] \\
& -\sum_{r,c,o,p,q,l,s,m}\left(\tau_{c,o,p,q,l,s,m}^{\text{land}}\cdot A_{r,t,c,o,l,s,m}\right) \\
& -\sum_{r}\left(\tau_r^{\text{live}}\cdot B_{r,t}\right)-\sum_{r,m}\left(\tau_{r,m}^{\text{proc}}\cdot P_{r,t,m}\right) \\
& -\sum_{r,\tilde{r},y}\left[\int \phi_{r,\tilde{r},t,y}^{\text{trad}}\left(T_{r,\tilde{r},t,y}\right)d\left(\cdot\right)\right].
\end{aligned}
\tag{1}
$$

Exogenous demand constraints:

$$
D_{r,t,y}\geq d_{r,t,y}^{\text{targ}}.
\tag{2}
$$

Product balance

$$
\begin{aligned}
D_{r,t,y}\leq & \sum_{c,o,p,q,l,s,m}\left(\alpha_{t,c,o,l,s,m,y}^{\text{land}}\cdot A_{r,t,c,o,l,s,m}\right)+\alpha_{r,t,y}^{\text{live}}\cdot B_{r,t} \\
& +\sum_{m}\left(\alpha_{r,m,y}^{\text{proc}}\cdot P_{r,t,m}\right)+\sum_{\tilde{r}}T_{\tilde{r},r,t,y}-\sum_{\tilde{r}}T_{r,\tilde{r},t,y}.
\end{aligned}
\tag{3}
$$

Land use balance

$$\sum_{s,m} A_{r,t,c,o,l,s,m} \leq L_{r,t,c,o,l}. \tag{4}$$

$$L_{r,t,c,o,l} \leq L^{\text{init}}_{r,t,c,o,l} + \sum_{\tilde{l}} Q_{r,t,c,o,\tilde{l},l} - \sum_{\tilde{l}} Q_{r,t,c,o,l,\tilde{l}}. \tag{5}$$

$$Q_{r,t,c,o,l,l} \leq L^{\text{suit}}_{r,t,c,o,l,l}. \tag{6}$$

Recursivity equations (calculated only once the model has been solved for a given period)

$$L^{\text{init}}_{r,t,c,o,l} = L^{\text{init}}_{r,t-1,c,o,l} + \sum_{\tilde{l}} Q_{r,t-1,c,o,\tilde{l},l} - \sum_{\tilde{l}} Q_{r,t-1,c,o,l,\tilde{l}}. \tag{7}$$

$$L^{\text{suit}}_{r,t,c,o,l,\tilde{l}} = L^{\text{suit}}_{r,t-1,c,o,l,\tilde{l}} + \sum_{\tilde{l}} Q_{r,t-1,c,o,\tilde{l},l} - \sum_{\tilde{l}} Q_{r,t-1,c,o,l,\tilde{l}}. \tag{8}$$

Irrigation water balance

$$\sum_{c,o,l,s,m} \left(\varpi_{c,l,s,m} \cdot A_{r,t,c,o,l,s,m} \right) \leq W_{r,t}. \tag{9}$$

Greenhouse gas emissions account

$$E_{r,t,e} = \sum_{c,o,l,s,m} \left(\varepsilon^{\text{land}}_{c,o,l,s,m,e} \cdot A_{r,t,c,o,l,s,m} \right) + \varepsilon^{\text{live}}_{r,e,t} \cdot B_{r,t}$$
$$+ \sum_{m} \left(\varepsilon^{\text{proc}}_{r,m,e} \cdot P_{r,t,m} \right) + \sum_{c,o,l,\tilde{l}} \left(\varepsilon^{\text{lucc}}_{c,o,l,\hat{l},e} \cdot Q_{r,t,c,o,l,\tilde{l}} \right). \tag{10}$$

Variables

D	demand quantity (tons, m³, kcal)
W	irrigation water consumption (m³)
Q	land use/cover change (ha)
A	land in different activities (ha)
B	livestock production (kcal)
P	processed quantity of primary input (tons, m³)
T	interregionally traded quantity (tons, m³, kcal)
E	greenhouse gas emissions (tCO₂eq)
L	available land (ha)

Functions

φ^{demd}	demand function (constant elasticity function)
φ^{splw}	water supply function (constant elasticity function)

φ^{lucc} land use/cover change cost function (linear function)

φ^{trad} trade cost function (constant elasticity function)

Parameters

τ^{land} land management cost except for water (\$/ha)

τ^{live} livestock production cost (\$/kcal)

τ^{proc} processing cost (\$/unit (t or m^3) of primary input)

d^{targ} exogenously given target demand (for example, biofuel targets; EJ, m^3, kcal)

α^{land} crop and tree yields (tons/ha, or m^3/ha)

α^{live} livestock technical coefficients (1 for livestock calories, negative number for feed requirements [t/kcal])

α^{proc} conversion coefficients (-1 for primary products, positive number for final products, for example, GJ/m^3)

L^{init} initial endowment of land of given land use/cover class (ha)

L^{suit} total area of land suitable for particular land uses/covers (ha)

ω irrigation water requirements (m^3/ha)

ε emission coefficients (tCO_2eq/unit of activity)

Indexes

r economic region (28 aggregated regions and individual countries)

t time period (10-year steps)

c country (203)

o simulation unit (defined at the intersection of 50 × 50 kilometer grid, homogeneous altitude class, slope class, and soil class)

l land cover/use type (cropland, grassland, managed forest, fast-growing tree plantations, pristine forest, other natural vegetation)

s species (37 crops, managed forests, fast-growing tree plantations)

m technologies: land use management (low input, high input, irrigated, subsistence, "current"); primary forest products transformation (sawn wood and wood pulp production); and bioenergy conversion (first-generation ethanol and biodiesel from sugarcane, corn, rapeseed, and soybeans; energy production from forest biomass—fermentation, gasification, and CHP)

y outputs (Primary: 30+ crops, sawlogs, pulpwood, other industrial logs, woodfuel, plantations biomass. Processed products: forest products (sawn wood and wood pulp), first-generation biofuels (ethanol and biodiesel), second-generation biofuels (ethanol and methanol), other bioenergy (power, heat, and gas)

e greenhouse gas accounts: CO_2 from land use change; CH_4 from enteric fermentation, rice production, and manure management; N_2O from synthetic fertilizers and from manure management; and CO_2 savings/ emissions from biofuels substituting fossil fuels

Table A.1 Input Data Used in the CongoBIOM Model

Parameter	Source	Year
Land characteristics	Skalsky et al. (2008), FAO, USGS, NASA, CRU UEA, JRC, IFPRI, IFA, WISE, etc.	
Soil classes	ISRIC	
Slope classes		
Altitude classses	SRTM 90m Digital Elevation Data (http://srtm.csi.cgiar.org)	
Country boundaries		
Aridity index	ICRAF, Zomer et al. (2008)	
Temperature threshold	European Centre for Medium Range Weather Forecasting (ECMWF)	
Protected area	FORAF	
Land cover	Global Land Cover (GLC 2000) Institute for Environment and Sustainability	2000
Agriculture		
Area		
Cropland area (1000 ha)	Global Land Cover (GLC 2000) Institute for Environment and Sustainability	2000
EPIC crop area (1000 ha)	IFPRI- You and Wood (2006)	2000
Cash crop area (1000 ha)	IFPRI- You et al. (2007)	2000
Irrigated area (1000 ha)	FAO	Average 1998–2002
Yield		
EPIC crop yield (T/ha)	BOKU, Erwin Schmid	
Cash crop yield (T/ha)	IFPRI- You et al. (2007)	2000
Average regional yield (T/ha)	FAO	Average 1998–2002
Input use		
Quantity of nitrogen (FTN) (kg/ha)	BOKU, Erwin Schmid	
Quantity of phosphorous (FTP)(kg/ha)	BOKU, Erwin Schmid	
Quantity of water (1000 m³/ha)	BOKU, Erwin Schmid	
Fertilizer application rates	IFA (1992)	
Fertilizer application rates	FAOSTAT	
Costs for 4 irrigation systems	Sauer et al. (2008)	
Production		
Crop production (1000 T)	FAO	Average 1998–2002
Livestock production	FAO	Average 1998–2002
Prices		
Crops (USD/T)	FAO	Average 1998–2002
Fertilizer price (USD/kg)	USDA (http://www.ers.usda.gov/Data/FertilizerUse/)	Average 2001–05
Forestry		
Area under concessions in Congo Basin (1000 ha)	FORAF	
Maximum share of sawlogs in the mean annual increment (m³/ha/year)	Kindermann et al. (2006)	
Harvestable wood for pulp production (m³/ha/year)	Kindermann et al. (2006)	
Mean annual increment (m³/ha/year)	Kindermann et al. (2008) based on the Global Forest Resources Assessment (FAO 2006a)	

table continues next page

Table A.1 Input Data Used in the CongoBIOM Model *(continued)*

Parameter	Source	Year
Biomass and wood production (m^3 or 1000 T)	FAO	2000
Harvesting costs	Kindermann et al. (2006)	
Short rotation plantation	Havlik et al. (2011)	
Suitable area (1000 ha)	Zomerat et al. (2008)	2010
Maximum annual increment (m^3/ha)	Alig et al. (2000); Chiba and Nagata (1987); FAO (2006b); Wadsworth (1997)	
Potential NPP	Cramer et al. (1999)	
Potentials for biomass plantations	Zomer et al. (2008)	
Sapling cost for manual planting	Carpentieri et al. (1993); Herzogbaum GmbH (2008)	
Labor requirements for plantation establishment	Jurvélius (1997)	
Average wages	ILO (2007)	
Unit cost of harvesting equipment and labor	FPP (1999); Jiroušek et al. (2007); Stokes et al. (1986); Wang et al. (2004)	
Slope factor	Hartsough et al. (2001)	
Ratio of mean PPP adjustment	Heston et al. (2006)	
GHG emissions		
N$_2$O emissions from application of synthetic fertilizers (kg CO$_2$/ha)	IPCC Guidelines (1996)	
Fertilizer application rates	IFA (1992)	
CO$_2$ savings/emission coefficients	CONCAWE/JRC/EUCAR (2007), Renewable Fuels Agency (2009)	
Above- and below-ground living biomass in forests (tCO$_2$eq/ha)	Kindermann et al. (2008)	
Above- and below-ground living biomass in grassland and other natural land (tCO$_2$eq/ha)	Ruesch and Gibbs (2008) (http://cdiac.ornl.gov/epubs/ndp/global_carbon/carbon_documentation.html)	
Total non-carbon emissions (million metric CO$_2$ equivalent)	EPA (2006)	
Crop carbon dioxide emissions (tons CO$_2$/hectare)	EPA (2006)	
GHG sequestration in SRP (tCO$_2$/ha)	Chiba and Nagata (1987)	
International Trade		
MacMap database	Bouet et al. (2005)	
BACI (based on COMTRADE)	Gaulier and Zignago (2009)	
International freight costs	Hummels et al. (2001)	
Infrastructure		
Existing infrastructure	WRI; Referentiel Geographique Commun	
Planned infrastructure	National statistics from Cameroon, Central African Republic, and Gabon and AICD (World Bank) for Democratic Republic of Congo, and Republic of Congo	
Process		
Conversion coefficients for sawn wood	4DSM model - Rametsteiner et al. (2007)	

table continues next page

Table A.1 Input Data Used in the CongoBIOM Model (continued)

Parameter	Source	Year
Conversion coefficients for wood pulp	4DSM model - Rametsteiner et al. (2007)	
Conversion coefficients and costs for energy	Biomass Technology Group (2005); Hamelinck and Faaij (2001); Leduc et al. (2008)	
Conversion coefficients and costs for ethanol	Hermann and Patel (2008)	
Conversion coefficients and costs for biodiesel	Haas et al. (2006)	
Production costs for sawn wood and wood pulp	Internal IIASA database and RISI database (http://www.risiinfo.com)	
Population		
Population per country (1,000 inhabitants)	Russ et al. (2007)	average 1999–2001
Estimated total population per region every 10 years between 2000 and 2100 (1,000 inhabitants)	GGI Scenario Database (2007)—Grubler et al. (2007)	
0.5 degree grid	GGI Scenario Database (2007)—Grubler et al. (2007)	
Population density	CIESIN (2005)	
Demand		
Initial food demand for crops (1000 T)	FBS data—FAO	average 1998–2002
Initial feed demand for crops (1000 T)	FBS data—FAO	average 1998–2002
Crop requirement per animal calories (T/1,000,000 kcal)	Supply Utilisation Accounts, FAOSTAT	average 1998–2002
Crop energy equivalent (kcal/T)	FBS data—FAO	
Relative change in consumption for meat, animal, vegetable, milk (kcal/capita)	FAO (2006a) World agriculture: toward 2030/2050 (Tables: 2.1, 2.7, 2.8)	
Own price elasticity	Seale, Regmi, and Bernstein (2003)	
GDP projections	GGI Scenario Database (2007)	
SUA data for crops (1,000 tons)	FAO	
FBS data	FAO	
Bioenergy projections	Russ et al. (2007)	
Biomass and wood consumption (m^3/ha or 1,000 T/ha)	FAO	

Databases

In order to enable global biophysical process modeling of agricultural and forest production, a comprehensive database—integrating information on soil type, climate, topography, land cover, and crop management—has been built (Skalsky et al. 2008). The data are available from various research institutes (NASA, JRC, FAO, USDA, IFPRI, etc.) and were harmonized into several common spatial resolution layers, including 5 and 30 arcmin as well as country layers. Consequently, Homogeneous Response Units (HRU) have been delineated by including only those parameters of landscape, which are almost constant over time. At the global scale, we have included five altitude classes, seven slope classes, and six soil classes. In a second step, the HRU layer is merged with other relevant information, such as a global climate map, land category/use map, irrigation map, and so on, which are actually inputs into the Environmental Policy Integrated Climate model (Williams 1995; Izaurralde et al. 2006). The Simulation Units are the intersection between country boundaries, 30 arcmin grid (50 × 50 kilometers), and Homogenous Response Unit.

Main Assumptions for the Baseline

Population growth: The regional population development is taken from the International Institute for Applied Systems Analysis (IIASA)'s SRES B2 scenario (Grübler et al. 2007). World population should increase from 6 billion in 2000 to 8 billion in 2030. In the Congo Basin, the model uses an average annual growth rate of 3.6 percent between 2000 and 2010 and 2.2 percent between 2020 and 2030, leading to a total population of 170 million people in 2030. The model uses the spatially explicit projections of population by 2010, 2020, and 2030 to represent the demand for woodfuel. No difference is made between rural and urban markets.

Exogenous constraints on food consumption: From the intermediate scenario of the SRES B2, GDP per capita is expected to grow at an average rate of 3 percent per year from 2000 to 2030 in the Congo Basin. FAO projections are used for per capita meat consumption. The model considers a minimum calorie intake per capita in each region and disallows large switches from one crop to another. The model currently restricts coffee and cocoa production to Sub-Saharan Africa. Initial demand for these crops is set at the observed imports in 2000 and is then adjusted for population growth. This assumption means that neither price changes nor income changes influence demand for coffee and cocoa.

Demand for energy: The model makes the assumption that woodfuel use per inhabitant remains constant, so that woodfuel demand increases proportionally to population. Bioenergy consumption comes from the POLES model (Russ et al. 2007) and assumes that there is no international trade in biofuels.

Other assumptions: The baseline is a situation where technical parameters remain identical to the 2000 level; new results are driven only by increases in food, wood, and bioenergy demand. There is no change in yields, annual increments, production costs, transportation costs, or trade policies. Subsistence farming is also fixed at its 2000 level. No environmental policies are implemented other than the 2000 protected areas. This baseline should be regarded as a "status quo" situation that allows us to isolate the impacts of various drivers of deforestation in the Congo Basin in the different scenarios.

References

Agritrade. 2009. "The Cocoa Sector in ACP-EU Trade." Executive brief, October 2009.

Andersen, P., and S. Shimorawa. 2007. "Rural Infrastructure and Agricultural Development." In *Rethinking Infrastructure for Development*. Annual World Bank Conference on Development Economics.

Angelsen, A., M. Brockhaus, M. Kanninen, E. Sills, W. D. Sunderlin, and S. Wertz-Kanounnikoff, eds. 2009. *Realising REDD+, National Strategy and Policy Options.* Bogor, Indonesia: CIFOR.

Atyi, R. E., D. Devers, C. de Wasseige, and F. Maisels. 2009. "Chapitre 1: Etat des forêts d'Afrique centrale: Synthèse sousrégionale." In *Etat de la Forêts 2008*, OFAC-COMIFAC.

Biomass Technology Group. 2005. *Handbook on Biomass Gasification.* H.A.M. Knoef. ISBN: 90-810068-1-9.

Bouet A., Y. Decreux, L. Fontagne, S. Jean, and D. Laborde. 2005. "A Consistent Ad-Valorem Equivalent Measure of Applied Protection Across the World: The MacMap HS6 Database. Document de travail CEPII N. 2004, December 22 (updated September 2005).

Carpentieri, A. E., E. D. Larson, and J. Woods. 1993. "Future Biomass-Based Electricity Supply in Northeast Brazil." *Biomass and Bioenergy* 4 (3): 149–73.

Chiba, S., and Y. Nagata, 1987. "Growth and Yield Estimates at Mountainous Forest Plantings of Improved Populus Maximowiczii." In *Research Report of Biomass Conversion Program. No.3 "High Yielding Technology for Mountainous Poplars by Short- or Mini-Rotation System I."* Japan; Agriculture, Forestry, and Fisheries Research Council Secretariat, Ministry of Agriculture, Forestry and Fisheries.

CIESIN (Center for International Earth Science Information Network), Columbia University; and Centro Internacional de Agricultura Tropical (CIAT). 2005. Gridded Population of the World Version 3 (GPWv3): Population Density Grids. Palisades, NY: Socioeconomic Data and Applications Center (SEDAC), Columbia University. http://sedac.ciesin.columbia.edu/gpw.

CONCAWE/JRC/EUCAR. 2007. "Well-to-Wheels Analysis of Future Automotive Fuels and Powertrains in the European context." *Well-to-Tank Report* Version 2c: 140.

EPA (Environmental Protection Agency). 2006. *Global Anthropogenic Non-CO_2 Greenhouse Gas Emissions: 1990–2020.* Washington, DC: United States Environmental Protection Agency.

FAO (Food and Agriculture Organization of the United Nations). 2006a. *Global Forest Resources Assessment 2005. Progress towards Sustainable Forest Management.* Rome, Italy: Food and Agriculture Organization of the United Nations.

———. 2006b. "Global Planted Forests Thematic Study: Results and Analysis." Planted Forests and Trees Working Paper 38, FAO, Rome.

Gaulier, G., and S. Zignago. 2009. "BACI: International Trade Database at the Product-level, the 1994–2007 Version." CEPII Working Paper.

Grübler, A., B. O'Neill, K. Riahi, V. Chirkov, A. Goujon, P. Kolp, I. Prommer, S. Scherbov, and E. Slentoe. 2007. "Regional, National, and Spatially Explicit Scenarios of Demographic and Economic Change Based on SRES." *Technological Forecasting and Social Change* 74: 980–1027.

Hamelinck, C. N., and A. P. C. Faaij. 2001. "Future Prospects for Production of Methanol and Hydrogen from Biomass." Utrecht University, Copernicus Institute, Science, Technology and Society, Utrecht, Netherlands.

Hass, M. J., A. J. McAloon, W. C. Yee, and T. A. Foglia. 2006. "A Process Model to Estimate Biodiesel Production Costs." *Bioresource Technology* 97:671–78.

Havlík, P., U. A. Schneider, E. Schmid, H. Boettcher, S. Fritz, R. Skalský, K. Aoki, S. de Cara, G. Kindermann, F. Kraxner, S. Leduc, I. McCallum, A. Mosnier, T. Sauer, and

M. Obersteiner. 2011. "Global Land-Use Implications of First and Second Generation Biofuel Targets." *Energy Policy* 39(10): 5690–702.

Herzogbaum GmbH. 2008. Forstpflanzen-Preisliste 2008. HERZOG.BAUM Samen & Pflanzen GmbH. Koaserbauerstr. 10, A-4810 Gmunden, Austria. Available at http://www.energiehoelzer.at.

Hummels D., J. Ishii, and K. M. Yi. 2001. "The Nature and Growth of Vertical Specialization in World Trade." *Journal of International Economics, Trade, and Wages* 54 (1): 75–96.

IFA (International Fertilizer Industry Association). 1992. *World Fertilizer Use Manual.* Germany: IFA.

IPCC (Intergovernmental Panel on Climate Change). 1996. "Revised 1996 IPCC Guidelines for National Greenhouse Gas Inventories." Intergovernmental Panel on Climate Change, United Nations Environment Programme, Organisation for Economic Co-Operation and Development, International Energy Agency, Paris.

Izaurralde, R. C., J. R. Williams, W. B. McGill, N. J. Rosenberg, and M. C. Q. Jakas. 2006. "Simulating Soil C Dynamics with EPIC: Model Description and Testing Against Long-Term Data." *Ecological Modelling* 192: 362–84.

Jurvélius, M. 1997. "Labor-Intensive Harvesting of Tree Plantations in the Southern Philippines." Forest Harvesting Case Study 9, RAP Publication: 1997/41, Food and Agriculture Organization of the United Nations, Bangkok, Thailand. http://www.fao.org/docrep/x5596e/x5596e00.HTM.

Kindermann, G. E., M. Obersteiner, E. Rametsteiner, and I. McCallum 2006. "Predicting the Deforestation-Trend under Different Carbon-Prices." *Carbon Balance and Management* 1: 15.

Kindermann, G., M. Obersteiner, E. Rametsteiner, and I. McCallum. 2008. "A Global Forest Growing Stock, Biomass and Carbon Map Based on FAO Statistics." *Silva Fennica* 42 (3): 387–96.

Leduc, S., D. Schwab, E. Dotzauer, E. Schmid, and M. Obersteiner. 2008. "Optimal Location of Wood Gasification Plants for Methanol Production with Heat Recovery." *International Journal of Energy Research* 32: 1080–91.

Rametsteiner, E., S. Nilsson, H. Böttcher, P. Havlik, F. Kraxner, S. Leduc, M. Obersteiner, F. Rydzak, U. Schneider, D. Schwab, and L. Willmore. 2007. "Study of the Effects of Globalization on the Economic Viability of EU Forestry." Final Report of the AGRI Tender Project: AGRI-G4-2006-06, EC Contract Number 30-CE-0097579/00-89.

Renewable Fuels Agency. 2009. "Carbon and Sustainability Reporting Within the Renewable Transport Fuel Obligation, Technical Guidance Part Two, Carbon Reporting—Default Values and Fuel Chains." Version 2.0, 207. Hastings, U.K.

Ruesch, A., and H. K. Gibbs. 2008. *New IPCC Tier-1 Global Biomass Carbon Map for the Year 2000.* Oak Ridge, TN: Oak Ridge National Laboratory. http://cdiac.ornl.gov.

Russ, P., T. Wiesenthal, D. van Regenmorter, and J. C. Císcar. 2007. *Global Climate Policy Scenarios for 2030 and Beyond—Analysis of Greenhouse Gas Emission Reduction Pathway Scenarios with the POLES and GEM-E3 models.* JRC Reference Reports. Seville, Spain: Joint Research Centre—Institute for Prospective Technological Studies.

Sauer, T., P. Havlík, G. Kindermann, and U. A. Schneider. 2008. "Agriculture, Population, Land and Water Scarcity in a Changing World—The Role of Irrigation." Congress of the European Association of Agricultural Economists, Gent, Belgium.

Seale, J., A. Regmi, and J. Bernstein. 2003. "International Evidence on Food Consumption Patterns." ERS/USDA Technical Bulletin No. 1904, Economic Research Service, U.S. Department of Agriculture, Washington, DC.

Skalsky, R., Z. Tarasovicova, J. Balkovic, E. Schmid, M. Fuchs, E. Moltchanova, G. Kindermann, P. Scholtz, et al. 2008. *GEO-BENE Global Database for Bio-Physical Modeling v.1.0—Concepts, Methodologies and Data.* The GEOBENE database report. Laxenburg, Austria: International Institute for Applied Analysis (IIASA).

Teravaninthorn, S., and G. Raballand. 2009. *Transport Prices and Costs in Africa: A Review of the Main International Corridors.* Washington, DC: World Bank.

Williams J. R. 1995. "The EPIC model." In *Computer Models of Watershed Hydrology,* ed. V. P. Singh, 909–1000. Highlands Ranch, CO: Water Resources Publications.

You, L., and S. Wood. 2006. "An Entropy Approach to Spatial Disaggregation of Agricultural Production." *Agricultural Systems* 90: 329–47.

You, L., S. Wood, U. Wood-Sichra, and J. Chamberlain. 2007. "Generating Plausible Crop Distribution Maps for Sub-Saharan Africa Using a Spatial Allocation Model." *Information Development* 23 (2–3): 151–59. International Food Policy Research Institute, Washington, DC.

Zomer, R. J., A. Trabucco, D. A. Bossio, and L. V. Verchot. 2008. "Climate Change Mitigation: A Spatial Analysis of Global Land Suitability for Clean Development Mechanism Afforestation and Reforestation." *Agriculture, Ecosystems and Environment* 126: 67–80.